Carpentry

ry

PRESTON'S COLLEGE

LEARNING ZONE

Tel: **01772 225310** Email: **learningzone@preston.ac.uk**

Please return or renew on or before the date
Fines will be charged on overdue items.

L

H

01 NOV 2006	15 JUN 2012	
24 JUN 2008		
09 OCT 2008		
09 OCT 2008		
27 MAR 2009		
22 MAI 2009		
12 JUN 2009		

D1493752

Class   694   MIT

Barcode

087947

THOMSON

Australia • Canada • Mexico • Singapore • Spain • United Kingdom • United States

THOMSON

Carpentry and Joinery

Copyright © George Mitchell 19982, 1995 and 1999

The Thomson logo is a registered trademark used herein under licence.

For more information, contact Thomson Learning, High Holborn House; 50-51 Bedford Row, London WC1R 4LR or visit us on the World Wide Web at:
http://www.thomsonlearning.co.uk

All rights reserved by Thomson Learning 2003. The text of this publication, or any part thereof, may not be reproduced or transmitted in any form or by any means, electronic or mechanical, including photocopying, recording, storage in an information retrieval system, or otherwise, without prior permission of the publisher.

While the publisher has taken all reasonable care in the preparation of this book the publisher makes no representation, express or implied, with regard to the accuracy of the information contained in this book and cannot accept any legal responsibility or liability for any errors or omissions from the book or the consequences thereof.

Products and services that are referred to in this book may be either trademarks and/or registered trademarks of their respective owners. The publisher and author/s make no claim to these trademarks.

British Library Cataloguing-in-Publication Data
A catalogue record for this book is available from the British Library

ISBN 1-84480-079-2

First published by Continuum in 1982
Second edition published by Continuum 1995
Reprinted 1998
Third edition published by Continuum 1999
Reprinted 2002 by Continuum
Reprinted 2003 by Thomson Learning

Printed in Singapore by Seng Lee Press

Acknowledgements

The author wishes to acknowledge, with grateful thanks, the help and assistance given by Mr C. Thompson in the preparation of this book. The author and publishers would also like to thank the following companies who kindly supplied information, illustrations and photographs for inclusion in this book: Record Ridgway Tools Ltd, Sheffield; Thomas Robinson and Son Ltd, Rochdale; Stanley Tools Ltd, Sheffield; Wolf Electric Tools Ltd, London; Myford Ltd, Beeston, Nottingham; Rentokil Ltd, East Grinstead, West Sussex; W. C. Youngman Ltd, Crawley, West Sussex; The Rawlplug Co. Ltd, Glasgow; Hitachi Power Tools Ltd, Milton Keynes; TRADA Technology, High Wycombe, Bucks.

Contents

Preface

In keeping with its two predecessors this third
edition of *Carpentry and Joinery* has been revised
and updated in order to satisfy the requirements
of all who use it and to extend its appeal to an
even wider readership.

Any person having an interest in the subject
matter, whether it be to prepare for an Institute
of Carpenters examination, an NVQ assessment
or simply as a hobby will find something of value
and interest in these pages. If used correctly, the
book will provide a foundation from which
access to the higher levels of craft competence
can be gained.

I gratefully acknowledge the assistance of David
Barker of Cassell in its preparation.

Dedicated to my wife, with sincere gratitude.

George Mitchell
January 1999

1 Timber technology and materials

The growth and structure of trees

To those studying botany a tree is just another plant. To the carpenter and joiner it is the plant which produces the material that forms the basis of the craft, namely *wood* (or timber). The growth of a tree is affected by the soil and the climate in which it grows. Soil has a considerable effect on the texture of timber, while climate influences the rate of growth which, in turn, affects the uses to which the timber can be put. Fast grown softwoods, for instance, are weaker and less suitable for use where strength is required.

The growth cycle is not the same in all trees and should be studied closely. To begin with, those trees which produce timber for commercial use are called *exogens* (outward growers). This means that the additional growth which occurs each year takes place on the outside of the stem (or trunk) just underneath its bark, while the innermost timber continues to mature. This period of annual growth begins in the spring and ends in late summer. Moisture is taken in from the soil by the roots and is passed up the stem (trunk) and through the sapwood area to the leaves. Having reached the leaves, and with the aid of the sun, this moisture is converted into sugars and starches (food) which nourish the tree as they are returned down the stem. This time, however, the route is different because the food travels between the *cambium layer* and the *bark* in an area called the *bast*. This cycle is shown in the diagram of tree growth.

The food is distributed to all parts of the tree by means of small cells which are called *rays*. These rays are more noticeable in the hardwoods than in softwoods, where they would usually only be seen through a microscope. Each time this cycle is completed the tree gains one more *growth ring* (also called an *annual ring*). It is by counting these rings that the age of a tree can be determined, as each ring represents one year of growth.

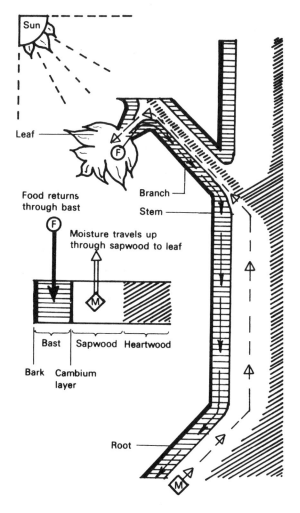

Diagram of tree growth

A *cross-section* showing structure is illustrated
and clearly shows what can be seen when a tree
is felled and viewed from its cross-cut section.
The *pith*, which is at the very centre, is of no
use after the tree has grown beyond the sapling
stage. Pith is easily recognised by its dark
brown colour and spongy texture. The use of
timber containing pith should be avoided.

Commercial timbers are classified into *softwoods*
and *hardwoods*. Probably the most common
method of identifying them is by their leaf.
Softwoods, which include pines, firs and
spruce, have a narrow needle-like leaf, whereas
hardwoods (oaks, beech and ash etc.) have a
broad leaf. The difference is more pronounced,
however, when the structure of both woods is
seen under a microscope. This close
examination reveals that softwoods are made up
of a series of rectangular cells called *tracheids*,
separated by walls through which the moisture
is passed. In the case of hardwoods nature has
devised a slightly different arrangement to
convey food to the various parts during the
trees' growth. This consists of a series of tube-
like ducts which are called *vessels*, or *pores*. In
hardwoods, such as oak and ash, the vessels are
reasonably regular and give a clear indication of
the growth ring. These are called *ring porous*,
but in many other hardwoods the vessels are
scattered and irregular and, for this reason, are
called *diffuse porous*.

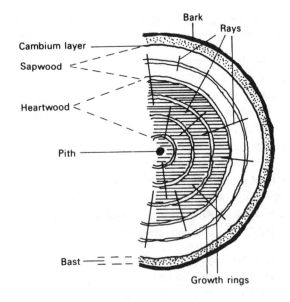

Cross-section showing structure

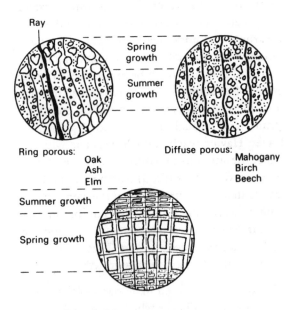

Softwood tracheids

Timber conversion

When a tree has matured it is cut down (felled) and sawn into timber sizes which are in common use in Industry. This process is called *conversion*. Fully mature trees should be felled in winter months for best results. The tree takes in least moisture during winter and, therefore, will begin to dry out (or shrink) at a slower rate.

The time taken for a tree to mature varies markedly between softwood and hardwood. The softwood tree will mature after 80 to 100 years while hardwoods take considerably longer at between 130 and 200 years. Any waste timber which occurs during conversion is recycled in the manufacture of certain building boards.

The way in which the log is cut will depend on the use to which the timber is to be put. For instance, general and structural timbers will be sawn in the *through and through* method which is shown. This method yields the maximum amount of usable wood from the log, but it must be remembered that timber cut in this way will also include growth defects which may result in poorer quality timber. If prime or first quality timber is required, the log will have to be *quarter sawn*. Two methods of quarter sawn conversion are illustrated, both of which will result in timber boards of good quality, but also with a fair amount of waste. This is one reason for the higher price of good quality timber. By the *tangential sawing* of logs it is possible to get boards with the maximum amount of grain configuration suitable for decorative work.

Because of the increasing scarcity of some of the finer hardwoods, such as *teak* and *mahogany*, there is a greater tendency to convert these timbers in the form of *veneers*. Further reference to this form of conversion is made under the manufacture of *plywood*. A veneer is the name given to a very thin slice of timber

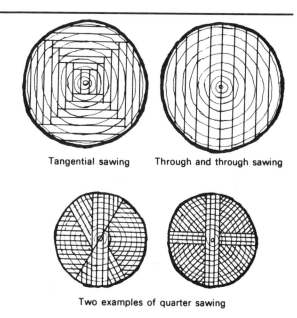

Tangential sawing   Through and through sawing

Two examples of quarter sawing

Methods of log cutting

which has a uniform thickness. This method enables much more face timber to be obtained from each log. The two most common ways of cutting a log into veneer is firstly by the rotary slicing method which is illustrated on page 23, and the second method is by cutting or slicing diametrically across the log (also shown on page 23). Whichever method is used the log is first softened up by a steam process. In addition to softening the log for easier slicing, this process also equalises the moisture contained in it. Only logs which are reasonably straight can be considered for use as veneers. Veneered boards are now extensively used in the cabinet making and furniture industries.

Moisture content and timber seasoning

When a tree is felled it contains a great deal of moisture. The timber will need to have some moisture content, whatever its commercial use is to be. However, this will need to suit the purpose and surroundings to which it is put and will be much less than is present when the tree is first cut down. The purpose of drying the timber is to minimise the subsequent movement that occurs when it is in use; which means that different uses demand different moisture content levels. Timber for internal use, for example, should have a lower moisture content than timber which is used externally. This is because high moisture levels in timber used internally would be reduced by the warmth of the atmosphere. This can cause excessive shrinking and possibly other more serious defects. A chart showing approximate levels of moisture content, suitable for the various types of use to which commercial timbers are put, is illustrated.

Moisture content is always expressed as a percentage of the dry weight of the timber. The formula used to calculate this percentage is:

$$\frac{\text{Wet weight} - \text{Dry weight}}{\text{Dry weight}} \times 100 = \text{Moisture content } (\%)$$

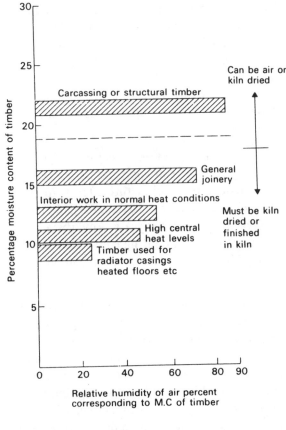

Moisture content levels for various uses

A sample of the timber is weighed accurately, this is the *wet weight*. It is then placed in a small kiln or oven at 100° Centigrade and taken out at intervals and weighed until no further loss of weight takes place. This last reading is called the *dry weight*. These two weights are then used as in the above formula to arrive at the moisture content of the timber sampled.

To meet the demands of industry this drying out process has to be carried out in large quantities. The process is called *seasoning*. The two most common methods of primary seasoning are *natural or air seasoning* and *kiln drying*. It will be appreciated that shrinkage in the timber must take place while boards are being dried and this means that the whole process must be controlled. *Degrade,* which is the name given to the lowering of quality, will occur if uneven or too rapid drying is permitted. A secondary method known as *second seasoning* is sometimes practised and will also be explained.

Natural or air seasoning

This process is the oldest method of drying timber and depends entirely upon a free flow of air to evaporate the moisture in the wood. The timber must be stacked correctly to obtain best results. There are three acknowledged disadvantages in the natural seasoning method:

1 It is difficult to control the rate of drying.
2 The timber dries at a slow rate, thus making the method uneconomical.
3 It is difficult to obtain a moisture content level less than 18–20% in natural conditions.

Various types of *drying shed* are used for this purpose, the most common consisting of steel stanchions supporting a roof over a solid concrete base. The timber stack will need to be protected from rain and other inclement weather, while in no way interrupting maximum ventilation.

Stacking the timber is of paramount importance. The boards should be stacked neatly with a gap of at least 25 mm between each. Piling sticks (sometimes called spacers) are placed between each layer of timber to enable air to get to all parts of the stack. These piling sticks vary in thickness in order to help control the drying rate. Sticks should also be of the same material as that being seasoned. For timbers where it is safe to dry fairly rapidly a piling stick of up to 38 mm in thickness can be used, but for timbers which must be dried slowly the gap between layers is reduced to only 12 mm. Piling sticks should be in vertical alignment and approximately 600 mm apart, although with thicker boards which are less likely to warp this spacing could be increased. Softwood stacks for natural seasoning should be no more than 2 m wide and 3 m high to allow a reasonable chance of uniform drying.
In larger stacks it is likely that the outer boards will dry much faster than those in the centre.

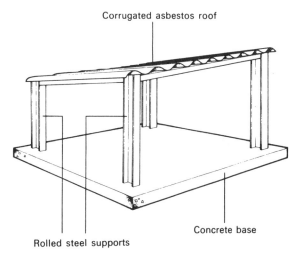

Corrugated asbestos roof

Rolled steel supports

Concrete base

Typical construction of a natural drying shed

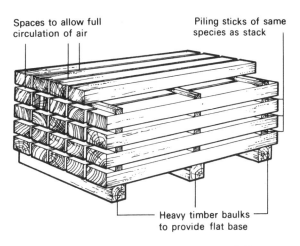

Spaces to allow full circulation of air

Piling sticks of same species as stack

Heavy timber baulks to provide flat base

Correct stacking method for softwoods (kiln or natural)

Hardwood stacking is rather different from the softwood method. It is more usual for the log to be converted *through and through* style and then stacked, again with sticks between, exactly as it would come off the saw. For best results, softwoods should be stacked in the spring whereas hardwoods should be stacked in the autumn to prevent the initial drying rate from being too rapid. Quite often, hardwoods are started off by natural methods and then finished off in the kiln. Softwood boards of 25 mm thickness reach 20% moisture content in approximately 3 months, while 50 mm boards require 4 months and perhaps more by this method. Hardwoods of 25 mm thickness would take between 7 – 8 months to reach 20% moisture content and 50 mm boards at least a

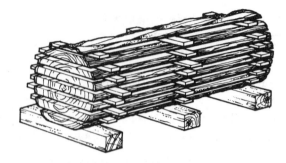

Hardwood log sawn and stacked for natural drying

year. These drying times can only be approximate because differences between timbers and weather conditions play an important part in seasoning by this method.

Kiln drying

This is by far the most common method of seasoning used today, mainly because of two important advantages:

1   It is capable of getting moisture content levels down well below that of natural drying. This often means that some timber started off in the drying shed is finished off in a kiln.
2   The fact that these levels can be reached safely in a controlled manner and in less time meets the demand of Industry much better.

Drying times in the kiln can be accurately estimated, but the instructions related to each species should be closely adhered to if degrades such as splits or warping are to be avoided.

There are various types of kiln, but most are built of brick. Inside the chamber there is a system of heating coils as well as steam jets. Most kilns will have fans to create circulation, but in a few cases natural draught kilns are used

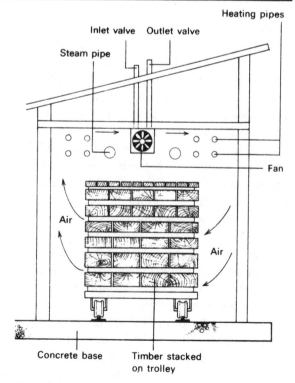

Forced draught type of kiln

in which circulation is maintained by natural draught ducts underneath the kiln.

Timber for drying must be stacked in exactly the same way as for natural seasoning, on trolleys running on a railway line which runs through the kiln. Once inside, the stack of timber is first subjected to an injection of steam to keep the humidity high inside the kiln. This action is followed by the gradual introduction of heat which circulates to all parts of the timber stack. As the heat increases the humidity is lowered and provided that the humidity and heat ratio is controlled throughout, no degrade should occur. In kilns which have forced draughts the fans can be put into reverse to ensure that uniform drying can be achieved. All this work is controlled from outside the kiln. Schedules prepared for the various timbers are followed closely. These are based on original and final moisture content readings. Kilns which have no forced draught facilities and rely upon natural draughts take longer, in some cases up to half as long again as those equipped with fans.

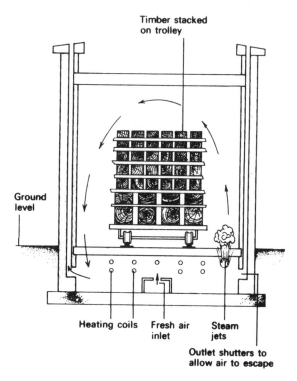

Natural draught type of kiln

Second seasoning

Second seasoning is the term used when seasoned timber is stored for a further period in the actual location where it is to be installed. This allows it to reach the exact moisture level of its surroundings, known as the equilibrium moisture content (EMC), but is usually only practised in the case of very high quality hardwood joinery. It is then customary for the joinery to be dry assembled only (e.g. not glued or fixed). After a short period, any re-fitting of joints or cleaning up necessitated by any further movement is carried out and the work is finally assembled and installed.

Timber defects

Defects which occur in timber and which have considerable effect on the usefulness of the timber fall into two categories:

1   Natural defects i.e. those which occur during growth, of which knots are the most common.
2   Those defects which occur after the tree has been felled. Many of the defects in this group could and should be avoided, particularly those resulting from carelessness during seasoning.

Natural defects

Knots occur in the tree when a branch is
formed. Trees which are forest grown have
fewer knots because of the confined space
around them and the general lack of sunlight.
The branch always starts at the centre of the
tree at the pith and therefore the resulting knot
shows in each board which is cut from the log.

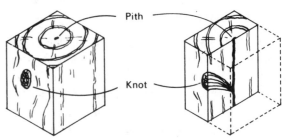

Left: Sound knot shown in a piece of timber
Right: The same piece cut through to show how the
       knot is connected with the pith

Sound knots

Sound knots will not fall out of the position
they occupy, but they tend to crack. This
provides inlets for any fungi to attack the wood.
They also deflect the grain of the plank and
this, too, is a weakness. Provided they are not
too large and not too near the edge of the
timber sound knots will not present a great
problem.

Sound knots

Dead knots

These are a source of real weakness, whatever
their size. They are produced when a branch is
broken off or removed before the tree has
finished growing. The tree reacts by starving
the part which is left inside the tree trunk,
causing it to die. This type of defect is easily
recognised by the very dark brown circle
around the decayed knot and this is liable to fall
out, eventually leaving a knot hole. Timber
containing dead knots is most unsuitable for
structural use and is normally classified as low
grade.

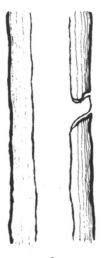

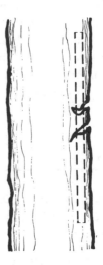

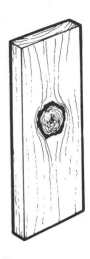

1	2	3	4
Branch broken off	Tree growth continues	Dotted line shows dead-knot area	Board containing a dead knot

Dead knots

Shakes (fissures) of various kinds are the other main natural defect in trees. They take the form of splits in the wood and detract from its strength quite considerably.

Heart shakes

These occur in the heartwood of a tree when it is left too long after it has matured before being felled for use. It is due to lack of food.

Ring shakes

Ring shakes are so called because they follow the contour of the growth ring. They usually result from excessive swaying of the tree in high winds, bringing about separation of the fibres.

Cup shakes

Cup shakes also position themselves in the growth ring and are brought about by similar conditions as ring shake, but without severe results.

Star shakes

This defect gives the appearance of a star following the lines of the rays. They are usually fine cracks and can be due to the sun drying up the cellular tissue when the bark has been damaged, or more often when the timber has been seasoned too quickly.

Upset is a form of shake, the cause of which is very uncertain. The effect of it is in the form of a zig-zag crack across the grain, which sometimes requires very close examination to be seen. It is a very serious defect because the affected boards are likely to snap very easily under very little pressure. The probable causes are:

1   The tree being struck by lightning during growth.
2   The tree falling awkwardly when felled, causing a fracture to run through the log.

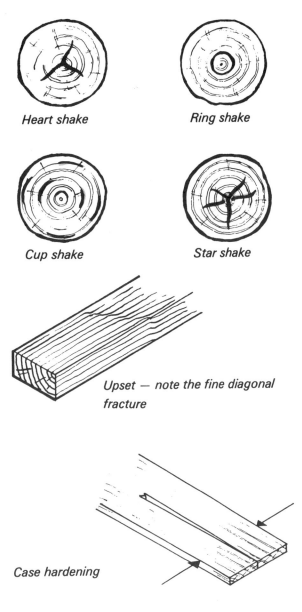

Heart shake

Ring shake

Cup shake

Star shake

Upset — note the fine diagonal fracture

Case hardening

Other defects

Most of the defects which occur after the tree has been felled are due to negligence of one kind or another.

Case hardening

This defect is the direct result of an imbalance in the humidity and heat ratio in the kiln. What actually happens is that the outer parts of the timber are dried while the centre or core is still green. This defect will cause the material to continue to move while in use. The operative will notice that when sawing a piece of such timber it will be inclined to pinch together before the cut is complete. A small timber spacer placed in the saw kerf (cut) will make cutting easier.

Honeycombing

Honeycombing occurs for the same reasons as case hardening, except that the hardened outer fibres apply such pressure on the moisture soaked inner fibres that numerous small holes can be seen on the end grain.

Honeycombing

Diamonding

This is the name given to the tendency of square cut pieces from certain areas of the log to become diamond shaped. It occurs when the piece has been cut with the growth rings running diagonally, causing the unequal shrinkage between summer and spring growth to pull it out of shape.

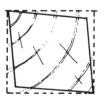

Diamonding

Cupping

Cupping will occur in the outer boards of a stack when moisture is permitted to evaporate faster on one face than the other. This is helped by the fact that shrinkage is greater on the outside (tangential face) of a growth ring than on the inside (radial face).

Cupping

Bowing

This often results when the boards are stacked with too much distance between the sticks.

Twisting

This defect is also known as *winding* and will occur when thin boards are cut from a log having curved longitudinal grain. The tendency is for the board to distort spirally.

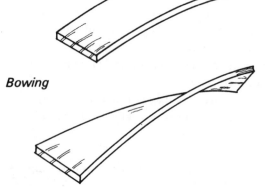

Bowing

Twisting

Timber decay

Decay is introduced into the timber in one of two ways:

1   Fungal attack
2   Insect attack

Both these forms of attack can be diagnosed and eradicated, but there are some conditions which occur during the growth of the tree which make it prone to attack by either fungi or insects when it is converted and brought into use. We shall look briefly at these conditions before further examining fungal and insect attacks on timber.

Druxiness is usually caused by water lying in the damaged section of bark after a branch has broken off. It is recognised as light brown blemishes in the timber.

Doatiness is a decay which is inherent in some hardwoods such as oak and beech. It is in the form of blackish spots and makes the timber very prone to dry rot.

Foxiness attacks hardwoods when they over-mature. It shows up as reddish-brown blotches.

This may be the first signs of dry rot

Fungal attack

The wood-destroying fungi which attack timber can be compared with plants because they feed off the wood in the same way as the plant feeds off the soil. Also, like the plant the fungus must have plenty of moisture to sustain it. This means that the moisture content level will need to be in excess of 20%. Timbers below this level are relatively safe. Conditions which favour the rapid growth of fungal attack are:

1   High moisture levels in the timbers
2   Little or no ventilation
3   Warm, humid atmosphere

Although oak and some other hardwoods have a fairly high resistance to fungal attack most timbers would break down if all three of the above elements were present.

Dry rot

Dry rot, or *Serpula* (formerly *Merulius*) *lacrymans* to give it its proper name, is so called because the timber eventually becomes dry and powdery. When the fungus has totally destroyed the area under attack its malignant strands will spread to other parts and continue to thrive. Areas most likely to be attacked in this way are enclosed areas such as underneath timber ground floors where its presence may not be known until it has reached an advanced stage resulting in the collapse of floorboards etc.

Eradication of dry rot must begin with the remedy of the cause, e.g. a blocked ventilator or broken water pipe and affected parts must be removed. These should be burnt for safety. All replacement timbers should be treated with a preservative, together with surrounding brickwork which may still contain live spores.

Cellar fungus

Cellar fungus or *Coniophora puteana* (formerly *cerebella*) is probably the second most common of the wood-destroying fungi and is frequently called wet rot. This, too, sets up in places with extremely wet conditions such as cellars, bathrooms and roofs. Dark brown strands are evident in the affected area, but, unlike dry rot, these are unable to spread to adjoining parts. Therefore, eradication is not so difficult. Only the immediately affected parts need be cut away and replaced with preserved timbers. Note, however, that the cause of the rot must also be remedied.

Lifting of floor boards may reveal an advanced stage of decay

Sap-stain

Sap-stain is more commonly called *blueing* because of the blue discoloration of sapwood that results. It is caused by storing freshly cut timber for too long a period in a warm atmosphere allowing the fungus to attack the walls of the cell, but not the cell itself. However, it should be noted that although it detracts from the appearance of the wood there is little reduction in its strength provided the wood has been thoroughly dried.

Insect attack

Insect attack on timber comes in various forms. There is a limited amount of damage caused by wood wasps and some members of the moth and butterfly families. By far the most devastating, however, are the beetles and it is upon these that we should focus our attention. The life cycle, briefly, is as follows:

1 Eggs are deposited in cracks or crevices of the timber.
2 The larva or grub which develops from the egg bores its way into the wood and creates the damage.
3 The grub develops into the pupa (dormant stage) until it finally emerges as an adult beetle.
4 The adult beetle bores its way out to commence the cycle in different places.

Death watch beetles are wood borers that prefer old, well seasoned soft or hardwoods on which

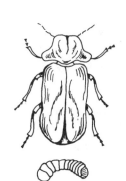

Death watch beetle and grub (4 × life size)

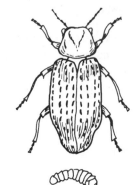

Common furniture beetle and grub (8 × life size)

Powder post or lyctus beetle and grub (5 × life size)

House longhorn beetle and grub (2 × life size)

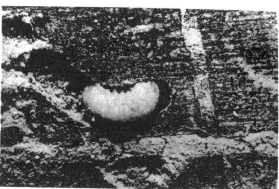

*The adult common furniture beetle
(20 × life size)*

The larva of the common furniture beetle

to thrive. Wood dust is their only food and in certain conditions their mating call, which is a tapping sound, can be heard. Its full name is *Xestobium rufovillosum*.

Common furniture beetle (*Anobium punctatum*) is a member of the same family as the death watch beetle. It was so named because it was once thought that it confined itself to attacking furniture. This is not so, they are not so fussy and can, in fact, create serious damage in structural timbers.

Powder post beetles (Lyctidae) are slightly different. They only attack hardwoods and, even more specifically, only hardwoods whose vessels are large enough to receive the eggs of the beetle. This is because the beetle is not a

prolific borer and the larva, when it becomes active, will only travel along the grain of the wood. Because of their small vessels and fine texture birch and beech are seldom attacked by this beetle.

House longhorn beetles (Cerambycidae) are the largest of this group and possess long antennae. This particular member of the longhorn group usually confines its attack to seasoned or partly seasoned sapwood in buildings.

Nowadays the control of these pests is by the use of insecticides, but a method of kiln sterilisation is available when large quantities of timber are under attack. In this process the timbers are placed in the kiln while steam and heat are used to suffocate the pests.

Timber preservation

From the previous section it will be understood that there is an urgent need to protect timber from fungal and insect attack. In addition, in cases where the timber may be exposed, it needs to be protected from the weather. Timber preservation is carried out for this purpose and the overall benefits from preservation can be summarised as follows:

Cost effective: by extending the life of the

timber, preservation is cheaper in the long term.

Maintenance: costs for this work will be considerably reduced or eliminated.

There are three groups of preservatives:

1   Tar-oil type
2   Water solution type
3   Organic solvent type

Chart showing suitability of preservatives for various uses.

Timber requirement	Coal-tar type	Water solution type	Organic solvent type
Fence posts Bridging Marine work Sleepers	Most suitable. Best results if pressure impregnated	Some are suitable. Avoid those which are known to leach	Considered too expensive for this work
Out-buildings Garages Fencing Weather boarding	Quite suitable even when hand applied	Suitable	Generally suitable
Roofing Structural Boat building General housing	Not considered suitable except in rare cases	Suitable	Suitable
Timber which may be close to or in contact with foodstuff. Timber used in enclosed conditions i.e. under floors	Not suited for any of these uses because of the strong vapour which is released	Suitable	Suitable

The best type of preservative to use in any given situation will vary according to where the timber is to be used.

Tar-oil

These preservatives are distilled from coal tar. *Creosote* is probably the best known. They are very efficient, but have a strong odour and should therefore not be used in confined areas or near foodstuffs.

Water solution

These types are nearly all odourless and can be painted over quite easily. The water is used as a vehicle to take the chemical into the timber and afterwards it evaporates, leaving the chemical to fight off attacks. Sodium fluoride, zinc chloride and copper sulphate solutions are commonly used as preservatives.

Organic solvents

This group consists of chemicals dissolved in volatile liquids such as white spirits. Many of this type are obtainable as proprietary brands of which 'Rentokil' is well known.

The application of these preservatives takes two forms:

1 Non pressure
2 Pressure impregnation

Non pressure application

There are disadvantages in all application methods other than pressure impregnation. The main disadvantages are:

1 The depth of impregnation is uneven and sometimes insufficient.

2   The various textures of the wood will sometimes not allow enough impregnation to prevent leaching (draining out) from taking place.

However, despite these disadvantages, it is better to apply the preservative and there are some instances where long and lasting benefits can be achieved.

Brushing is probably the most common method of application. The preservative must be applied liberally (not brushed out) to allow it to soak in.

Spraying is used in areas difficult to get into, such as roof spaces. Protective clothing should be worn.

Dipping can produce fairly good results, especially on timbers with open vessels. The timbers are submerged in a bath of preserving liquid for between 5 and 15 minutes. Excess preservative is allowed to drain into the bath or tank.

Steeping is similar to dipping except that the timbers are left submerged for as long as possible i.e. up to two weeks. This method is more necessary with the water solution types of preservative because of their slower rate of impregnation.

Note: Because all these liquids are toxic and some are likely to burn, care should be taken at all times when using them. Protective clothing should be worn when large quantities are being treated.

Pressure impregnation
There are two principal methods of pressure impregnation:

1   Full cell process
2   Empty cell process

The use to which the timber is to be put will usually determine the process required.

Brush application must be liberal to achieve penetration

Spraying roof timbers — note use of protective mask

A batch of boarding about to be dipped in a tank of preservative

In the *full cell* method the timber is placed in a cylinder and the doors are sealed. Prior to impregnation a vacuum is created inside the chamber and then the preservative is introduced at a temperature of about 150°F. When the cylinder is completely filled with liquid, pressure is applied to force the liquid into the cells of the timber. The cells are filled to capacity.

The *empty cell* process differs from this in that after impregnation the treated wood is relieved of all surplus liquid by compressed air. This leaves the cells empty, but the walls of the cells are fully preserved.

The entire process is carried out in plant specially designed for the purpose. The main chamber is cylindrical in shape with a narrow gauge railway line passing through on which trolleys laden with timber can be moved in and out. Outside this cylinder, but close by, will be situated a storage tank for the chemical preservative and a mixing tank. A pump to provide the pressure is placed just outside the cylinder, in a convenient position.

A vacuum pressure impregnation plant

The photograph shows the depth of impregnation that can be achieved by the vacuum pressure process. Note that the whole of the sapwood area is treated.

The plant in use

The depths of impregnation achieved

Softwoods and hardwoods in common use

Softwoods

Most of the softwoods now used in the building and construction industry are conifers. This means that they are cone bearing trees belonging to the Pinaceae family.

Douglas fir

This is probably the largest and strongest of all the softwoods. It is reddish-brown in colour and has good resistance to decay and acids. These qualities are the reason why most of the shuttering plywoods are made from it. In addition, it is extensively used in the construction of timber framed buildings as well as laminated structures. Large forests of the timber are grown in North America.

European redwood

This softwood is used extensively in both carpentry and joinery work. It is also called red or yellow deal and is forested in Europe, Northern Asia and Scotland, where it is known as Scots pine. The timber is strong and fairly hard with a pale reddy-brown appearance. Although it tends to have numerous knots it works well and is in plentiful supply.

Parana pine

Parana pine has not been imported into the UK for as long as other well known softwoods. There are dense forests of the tree in Brazil and other parts of South America. It is obtainable in very wide and long boards which makes it useful for large work. However, it is very prone to twist and bow. The wood is very hard and strong. Its colour varies quite considerably from light to dark brown with bright red streaks present in some boards. Its uses include internal fittings and joinery and it works well with hand or machine tools.

Pitch pine

Pitch pine grows in the Gulf States of the USA.
It is very resinous and provides most of the world's resin supplies. Church work and high grade joinery are some of its uses, but because of its size and strength, it is also used on large structural work. Creamy light-brown in colour with definite grain markings, it responds well to glue and can be nailed without splitting.

Spruce

This is a whitish timber having fine brown lines. It comes from North America and Northern Europe and is used for interior joinery work. Because of its light weight the timber is not considered very strong. It takes glue and nails well and generally works easily. Supplies are reasonably good.

Western red cedar

This is one of the largest trees when fully grown, being around 5 m in diameter and 60 m high. It seasons well and is immune from fungal attack and also has a high resistance to insect attack, which is why it is frequently used externally. Although it is not a strong timber its durability is good. Its colour ranges from light to dark-brown and it is used extensively in timber buildings of all types.

Hardwoods

There are very many relatively unknown hardwoods used in joinery and interior design at the present time. Only those which are in common use are included here.

Beech

Beech is a very hard, close grained timber which is grown in Central and Southern Europe. The colour ranges from reddish-brown to light-brown and is easily recognised by its speckled face. It is used a great deal in the furniture industry and works well with sharp tools.

Birch

This is also from Europe, although there are forests in Canada and North America. This wood has a very fine texture and is off-white to creamy-brown in colour. It is not a large tree and is therefore extensively used for veneers in the plywood making industry. This species of timber is inclined to warp and twist, but is very hard.

Iroko

Iroko resembles teak in appearance, but has not got its fine qualities. The tree grows right across the African continent and is sometimes wrongly referred to as teak. It works well and is very strong and durable. Colours vary from yellowish-brown to dark-brown depending upon where it is grown. It is used for good class joinery and because of its high resistance to moisture is also suitable for dock work.

Mahogany (African)

This hardwood grows in West Africa and Cuba. It is the most widely used mahogany nowadays, due to the fact that it is in good supply. It is difficult to work due to grain variations, but is quite strong and resistant to decay. The main uses for this timber include panelling and internal joinery as well as for veneers. Its colour varies from deepish red to light-brown.

Oak

To many people oak is the supreme hardwood. It is certainly a hard, nicely figured timber. Fine tools are required to achieve a high quality finish. The colour is usually biscuit shade with a silvery fleck. Supplies are obtained from Europe, North America and Japan and it is usually reserved for good quality joinery, furniture or panelling. It responds well to glue. Most species, however, are difficult to season, as they have a tendency to split. For this reason seasoning is usually started naturally and then finished by kiln.

Obeche

This wood comes from West Africa and the Gold Coast. The timber is creamy-white to pale-yellow in colour with an interlocking grain which gives a faint stripe to its appearance. Because it is not very durable it is used for internal joinery, but it takes glue and nails well.

Ramin

Ramin is a whitish timber with light-brown streaks and mild texture. It is a stable wood with good strength compared to its weight. Ramin is widely used in the manufacture of small and large diameter dowelling as well as beading and mouldings. It works well, and takes glue and nails easily. It is grown in Malaysia.

Teak

This is another hardwood with very fine qualities. The best quality teak is imported from Burma and India with other species coming from Indo-China. It is an extremely durable and strong timber with a high resistance to moisture, fire and acids. Teak works well with good quality hand or machine tools and a high quality finish is obtainable. Rich brown in colour it sometimes has a grain figure, but not often. It has a wide variety of uses in furniture, panelling, doors, frames and is extensively used for external woodwork. Unfortunately, it is becoming scarce.

Utile

Utile has been used in increasing amounts in recent years to replace the diminishing supplies of mahogany, which it vaguely resembles. It is a deep reddish colour with paler sapwood and occasionally shows attractive figuring. The tree grows in large quantities in both East and West Africa and is used for cabinet making and joinery. It is easy to work with sharp tools and takes glue very well. It has strength proportionate to its weight.

Manufactured boards

A range of sheet materials are now widely used in the woodworking and building industries. Some are made from solid timber while others are made from what would otherwise be waste wood. This is the material referred to earlier as being of such low grade that it could not be used in the ordinary sense. The standard size for all building boards is 1220 × 2440 mm, but boards of larger and smaller size are obtainable. The thickness of the boards vary according to their use.

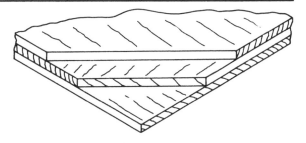

Position of veneers in 3 ply

Plywood

Three-ply plywood consists of three veneers of equal thickness glued together with the centre, or core veneer, having its grain running at right angles to the outer veneers. This gives the board considerable strength while making it easy to conform to circular or other shapes. It has extensive uses which include panelling and linings as well as other forms of covering.

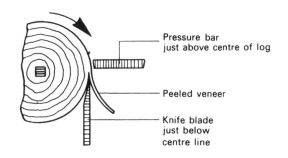

Rotary or log peeling veneer production

The grade is usually stamped on the board by the manufacturer and these must be recognised when deciding the use to which the board is to be put. For instance, INT denotes that it is suitable for *internal* use only and indicates that it has been glued together with glue having low moisture resistance. If such a plywood is used in situations exposed or subject to moist conditions there would be a separation of the veneers in due course. Modern adhesives make it possible for varying grades to be produced. MR indicates *fair resistance to moisture,* BR means that the ply has a fairly high resistance in exposed conditions because it is *boil resistant* and WBP indicates that it is *weather and boil proof.* This last type is used in most adverse conditions and also in the boat building industry.

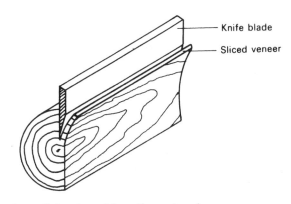

Log slicing to achieve figured grain

The veneers are mainly produced by rotary cutting. Plywood is sometimes made with one hardwood veneered face. This veneer is extra to those used in making the plywood and is often tangentially cut to obtain maximum figure on the face. Whenever this is done a balancing veneer, usually of lower grade, must be applied to the reverse side to maintain balance unless double faced boards are needed. Many timbers are used in the manufacture of plywood and

birch is one of the more common. It produces a
good, stable board. Three-ply thicknesses range
between 3 and 5 mm, after which it becomes
multi-ply.

Multi-ply

This is a board containing five or more veneers
glued together, as in three ply, with grains
running at 90° to each other. There should
always be an odd number of veneers if the
balance of the sheet is to be maintained. Multi-
ply is graded as for three-ply and is also
obtainable with face veneers. The thickness of
multi-ply boards ranges from 6 mm up to 24
mm. Douglas fir sheathing is a multi-ply of
WBP grade which is used extensively in
formwork because of its ability to be re-used
many times. Birch is often used as a face veneer
for multi-ply boards.

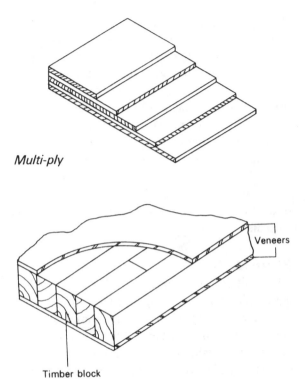

Multi-ply

Blockboard

Blockboards

This is another type of built-up board, but this
time the core is made up of strips of solid
timber up to 25 mm wide. The strips are glued
together with thick veneers on both outer faces.
This produces a fairly stable and very strong
board which has very many uses. Blockboards
with a low grade timber core are quite likely to
twist. Once again good quality hardwood
veneers are sometimes applied.

Laminboard

Laminboard provides an even better quality
than blockboard, because the core strips are
much narrower and never exceed 7 – 8 mm.
This adds stability to the board and minimises
the amount of poor quality timber used in the
core.

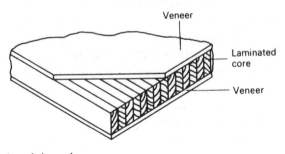

Laminboard

Battenboard

Battenboard uses a core strip much wider than
either block or laminboard, sometimes as much
as 75 mm wide. This results in the board being
of poorer quality because it is not so able to
keep its shape.

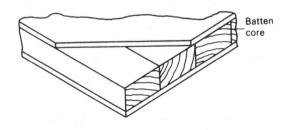

Battenboard

Chipboard

Chipboard consists of compressed wood shavings bonded with synthetic resin glue. Thicknesses range from 13 mm to 22 mm, but there are various grades. For instance, a special flooring quality is manufactured tough enough to replace normal solid timber flooring boards. The main disadvantage with chipboard is that it will not take a screw. However, it is extensively used for the cheaper range of furniture as well as for wall coverings etc. It is also produced with hardwood veneer finishes and plastic coatings.

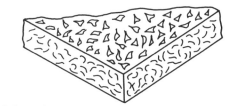

Chipboard

Hardboard

Hardboard is yet another of the wood waste-products. The wood is pulped by machine, bonded with adhesive and finally pressed to thicknesses of 3 mm to 6 mm. Hardboard has very many uses including panelling, wall and floor covering. The surface is very tough and can be difficult to penetrate with panel pins. For best results all hardboard should be wetted on its back side and allowed to dry before use. This will prevent any distortion after the sheet has been finally fixed. Reeded and fluted hardboards and some with other face features are obtainable for decorative work. Pre-finished boards are also widely used.

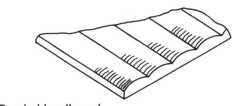

Reeded hardboard

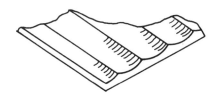

Fluted hardboard

Insulating boards

These boards are used extensively in sound and thermal insulation. They are made in thicknesses which range from 6 mm to 25 mm. The boards are much softer than hardboard and are made from pulped paper and wood. They are not, however, subjected to anything like so much pressure in manufacture. This gives them a spongy, absorbent texture.

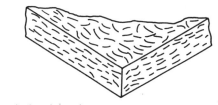

Insulation board

Medium density fibreboard (MDF)

MDF is made from wood fibres bonded together with synthetic resin glues (described under *Adhesives*) and is being used extensively in Joinery manufacture as well as applied mouldings such as skirtings, architraves etc. It can be machine-routed and finished to any desired shape or section and is easy to work with conventional tools. MDF is made to standard sheet sizes with thicknesses ranging from 6 mm to 30 mm. A moisture-resistant MDF is also available for external use.

Sterling board (Oriented Strand Board OSB)

This is now widely used as a less expensive alternative to exterior-grade plywood. It is made up of wood strands glued together in such a way as to produce great strength in the boards as well as rigidity. Sheet sizes are standard with thicknesses ranging from 6 mm to 25 mm.

Note: It must be remembered when using *any* particle board that they all cause a certain amount of dust which some will find an irritant. It is wise, therefore, to wear a mask at all times. Furthermore, it is important to stack all sheet materials in such a position as to ensure that they will remain flat and free from damage.

Questions

1 Describe the annual growth cycle of a tree.

2 What does a growth ring denote?

3 Describe the main differences in the structure of hardwood and softwoods.

4 Why should trees be felled during the winter?

5 Which method of converting a log produces the best quality timber?

6 Why is it necessary to season timber to different levels of moisture content?

7 List the advantages and disadvantages of natural and kiln drying.

8 (a) List the types of decay that can affect timber.
   (b) At what percentage moisture content is timber most likely to be attacked by dry rot?

9 Describe the life cycle of a wood-boring beetle.

10 List and describe the main groups of timber preservative.

11 What is the standard metric size of sheet material?

12 Why are various adhesives used in ply and blockboard manufacture?

2 Tools—their selection and use

Woodwork is one of the most ancient of crafts, yet it is surprising how many of the tools still in use today are updated versions of those used by the early woodworker. Modern materials and production methods have made it possible to produce tools of good and lasting quality. Those whose main work will be in the craft would be wise to use only those tools made by firms with a good reputation.

Marking and setting-out tools

Whether on site work or in the joinery shop there is a good deal of *marking-out* and *setting-out* that the woodworker is responsible for. Only accurate and appropriate tools should be used for this purpose. *One metre rules* are extensively used in both shop and site work. They can be obtained in flexible plastic capable of withstanding rough usage when necessary. The rule is marked out in 5 mm units with every 10 mm (1 cm) being shown clearly for instant recognition. Each 100 mm is also clearly indicated up to the 1 m mark. It may be found more convenient to have the type of rule which is marked in both imperial and metric units.

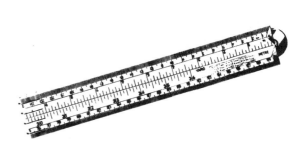

Durable rule with imperial and metric units

Spring tapes are also widely use for measurements over 1 m. These are available up to 5 m in length. The better quality tapes have both a locking device to retain the tape in the open position during use and a return spring which will retract the tape speedily when it is no longer required. Site measurement can be quicker and more accurate with this type of rule.

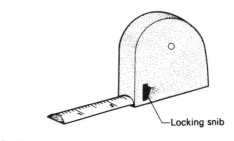

—Locking snib

Spring tape

The *marking knife* is used in joinery marking-out to give greater accuracy and better finishes to shoulder lines and lines across the grain etc. Many have a steel blade with hardwood handle, but some all steel knives are available.

Try squares of varying sizes with steel blades and hardwood stocks are also widely used. It is of paramount importance that these are accurate and a method of checking this accuracy is illustrated.

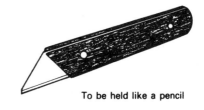

To be held like a pencil

Marking Knife

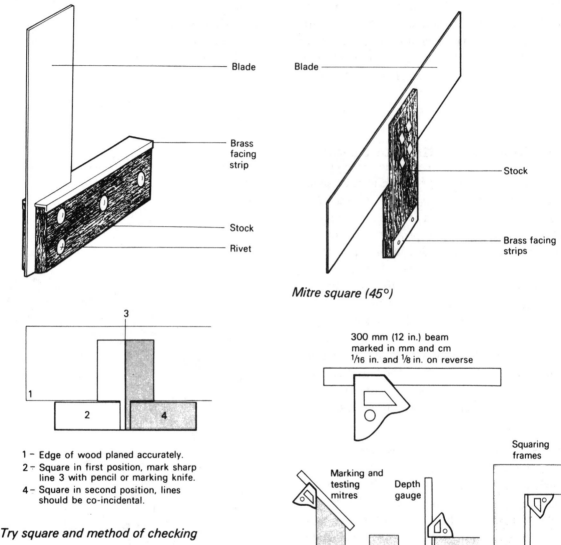

Mitre square (45°)

1 – Edge of wood planed accurately.
2 – Square in first position, mark sharp
     line 3 with pencil or marking knife.
4 – Square in second position, lines
     should be co-incidental.

*Try square and method of checking
accuracy*

300 mm (12 in.) beam
marked in mm and cm
1/16 in. and 1/8 in. on reverse

Squaring
frames

Marking and
testing
mitres

Depth
gauge

Multi-purpose square and some of its uses

Mitre squares, which consist of a steel blade set at 45° to the hardwood stock, are also very useful.

Multi-purpose squares are all metal and combine both the try square and mitre square in one tool. They are valuable on site work for various small operations, some of which are illustrated.

These squares have a 300 mm beam which can be set in any position required.

Adjustable sliding bevels have slotted steel blades with either steel or hardwood stocks. They are used for marking all angles greater or smaller than 45°. The adjustment is obtained by releasing a screw or a lever at the top of the

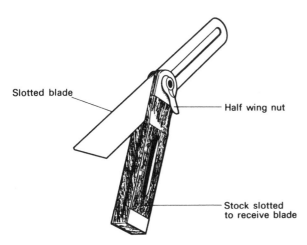

Adjustable sliding bevel

Slotted blade

Half wing nut

Stock slotted
to receive blade

stock and, having set the required angle, tightening up the screw or lever. Care should be taken to avoid overtightening these tools when setting.

The marking gauge consists of a hardwood stem which has a steel spur at one end and a hardwood stock. A plastic or hardwood set screw retains the stock in the required position. The tool is used in all accurate work and gives a fine cut line on the timber which the carpenter can work to. It should be used with the spur in the trailing position as shown.

Mortise gauges have a double spur for denoting the position of mortises and grooves. One spur is fixed while the other is adjusted by a threaded rod and thumb screw which is located at the end of the stem. Some mortise gauges also have a single spur on the opposite side of the stem to combine both marking and mortise gauge. The spurs should first be set to the required mortise or groove size followed by the positioning of the stock with a rule.

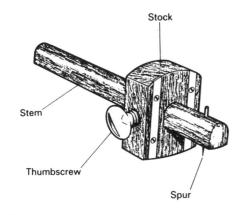

Stock

Stem

Thumbscrew

Spur

Marking gauge

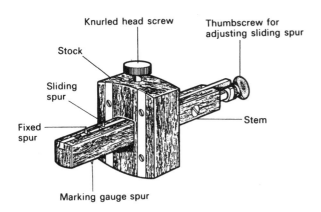

Knurled head screw

Thumbscrew for
adjusting sliding spur

Stock

Sliding
spur

Fixed
spur

Stem

Marking gauge spur

Mortise gauge

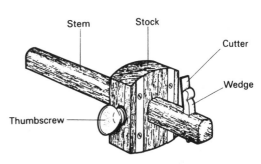

Stem

Stock

Cutter

Wedge

Thumbscrew

The blade can be ground with
a curve (1 and 2) or to give
a diamond point (3)

1   2   3

Cutting gauge

Cutting gauges are used mainly on work of great accuracy and good quality. They differ from the marking gauge only in the fact that they have a small cutter which is sharpened to a chisel edge instead of a spur. The cutters can be shaped to suit various requirements. They are held in position by a brass wedge to enable the cutter to be adjusted in depth if necessary. These gauges are capable of forming small rebates or other sorts of recess if used carefully.

Saws

Saws of one kind or another are among those tools most frequently used by woodworkers. Some confusion has often surrounded the term *handsaw*. This term can be applied to three different members of the saw family which are:

1 Rip saw
2 Cross-cut saw
3 Panel saw

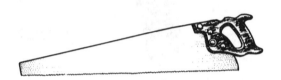

Handsaw

The differences between these three types of saw is the length of the blade and also the number of teeth per 25 mm of blade. Those saws with few teeth per 25 mm will have a coarse action, while those with a greater number will have a fine cut. The number of teeth will also determine the amount of *set* given to each tooth. The set is the alternate bending of the teeth with a special tool to give clearance in the kerf (cut). All saws cut wood by the action of their teeth severing the wood fibres.

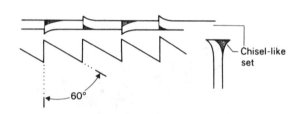

Ripsaw teeth — for cutting with the grain only

Rip saws

Rip saws are the largest and coarsest of this group. Their length is between 650–710 mm and the number of teeth per 25 mm is four and a half or five. This type of saw is used specifically for ripping timber along its grain and for this reason it is sharpened differently to other saws. The teeth are filed square across the saw to give a square chisel-like tip to each tooth. In addition, a large set is given to each tooth to provide ample clearance.

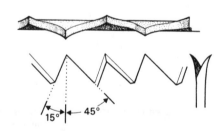

Cross-cut and panel saw teeth

Cross-cut saws

Cross-cut saws are slightly smaller in length at 600–660 mm and have between seven and nine teeth per 25 mm. As the name suggests its main use is cutting across the grain, but it will perform some ripping work provided it is not strained.

Panel saws

Panel saws are the finest of the handsaws. Their

length is usually between 550 and 610 mm and they possess ten to twelve teeth per 25 mm. This saw is exceptionally useful and versatile. However, the craftsman needs to take care not to extend it beyond its capacity.

Tenon saws

Tenon saws are the largest of the backsaws. The name derives from the steel or brass back which is fitted to the back of the saw to maintain accuracy and prevent buckle. The saw itself is 250 – 350 mm long with twelve to fourteen teeth per 25 mm. It will cut small tenons very efficiently and carry out many other difficult sawing operations where accuracy is essential.

Tenon / dovetail saw

Dovetail saws

The dovetail saw is a smaller and finer version of the tenon saw. They are invaluable for dovetail work and other very fine operations. Their length is in the range of 200 – 255 mm and they have 14 plus teeth to every 25 mm. Care must be taken to keep the saw in first class condition for best results.

Coping saw

The coping saw is a very versatile frame saw used in the cutting of shaped work and for cutting away waste material when forming other joints. The small steel frame holds the fine blade in position and the threaded handle maintains the tension when screwed in fully. The blade should be set so that the cut is made on the forward stroke and the frame can be positioned at any angle by slackening the handle.

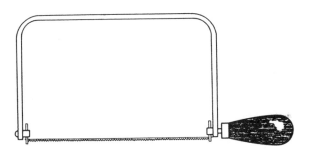

Coping saw

Bow saw

Bow saws are confined to joinery shop work nowadays because they are too large to carry about in the tool box and the coping saw does much the same work anyway. This saw consists of a hardwood frame with the blade held taut by

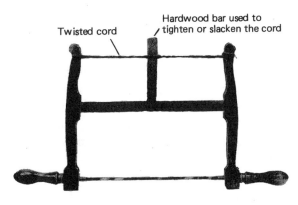

Twisted cord

Hardwood bar used to tighten or slacken the cord

Bow saw

a twisted cord stretched across the top of the
frame in the form of a tourniquet. Once again,
the primary use is for shaped and curved work.

Pad saw

Pad saws are mainly used for shaped work, but
are also useful for straight cutting when boards,
doors or panels require apertures to be cut. The
two types available are the traditional model
consisting of a wooden handle through which
the blade passes and is secured by two threaded
bolts in the ferrule. The more modern type has
a plastic or alloy handle through which the
blade passes and is secured by a knurled thumb
screw located underneath the handle near the
front.

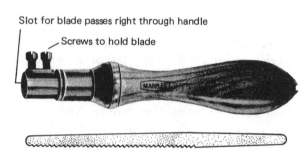

Slot for blade passes right through handle

Screws to hold blade

Traditional pad handle and blade

Junior hacksaw

The junior hacksaw, although not strictly a
woodworking tool, is a very useful part of the
carpenter's kit. It basically consists of a simple
metal frame into which the blade is sprung.
They occupy very little tool-box space, but are
invaluable for cutting off metal lock spindles,
bolt lengths and other small metal parts.

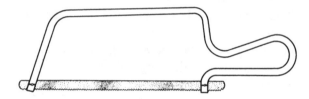

Junior hacksaw

Planes

Metal planes have now taken the place of the
traditional wooden ones, but the functions they
perform are the same. *Trying planes, jack planes*
and *smoothing planes* are made identically, the
only difference being the size of the plane and
the use to which it is put. They all have
precision ground soles to which are attached the
handle at the back and a grip knob at the front
to control the tool. In a central position is the
part called the *frog* which has an angle of
approximately 45°. The cutting iron assembly
consists of a cutting iron to which is bolted a
cap iron to break wood shavings and give
support to the cutting iron for a better cutting
action. The complete assembly is retained in

position by the part known as the *lever cap*.
Lateral movement can be obtained by means of
a lever found at the upper end of the frog. The
amount of cut is controlled by a brass knurled
turn-screw which is situated behind the frog
about a finger length away from the handle.

The *trying plane* is by far the largest of this
group, being 560 mm in length with a cutting
iron width of 65 mm. Its use is confined to
getting straight and true surfaces on boards
such as when two or more boards are to be
jointed together. For this reason the plane is
also called a *jointer*.

Jack planes on the other hand are very

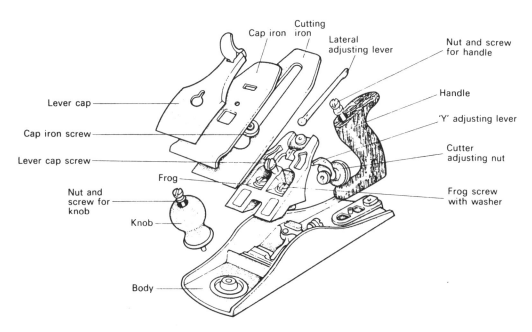

Cap iron

Cutting iron

Lateral adjusting lever

Nut and screw for handle

Lever cap

Cap iron screw

Lever cap screw

Frog

Nut and screw for knob

Knob

Body

Handle

'Y' adjusting lever

Cutter adjusting nut

Frog screw with washer

The parts of the smoothing, jack and try plane

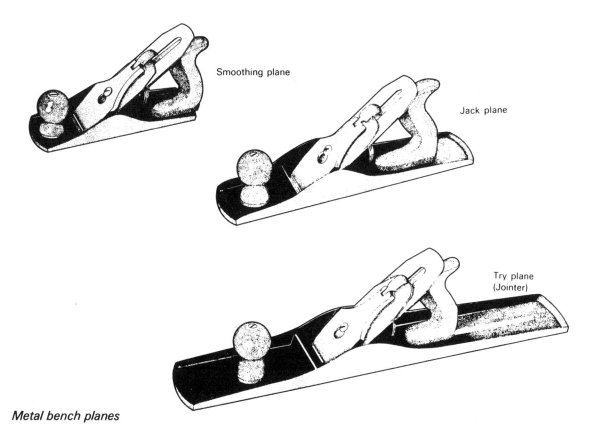

Smoothing plane

Jack plane

Try plane (Jointer)

Metal bench planes

extensively used. Their length is approximately 380 mm and cutting iron widths can be as small as 50 mm or as large as 62 mm. The many uses to which they are put includes removal of waste, shooting door edges accurately and putting chamfers and bevels on timber as required, ('shoot' or 'shooting' are trade terms meaning 'to create a straight edge with a plane'). The cutting iron should have a very slightly curved cutting edge and the cap iron should be no closer than 2 mm to the cutting edge.

Smoothing planes are usually 260 mm long with iron widths of 50 or 62 mm. They are primarily for finishing work and for this reason a straight cutting iron edge is best. In addition the distance between the cap iron and cutting edge should be between 1 and 1.5 mm for fine, accurate work.

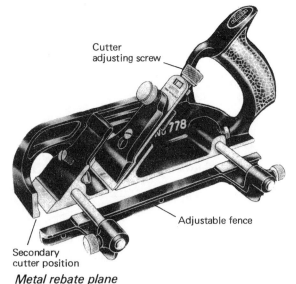

Metal rebate plane

Rebate planes are of different design, although they also have a precision-ground sole with a side return precision-ground at 90° to the sole for accurate rebating. The cutting iron is

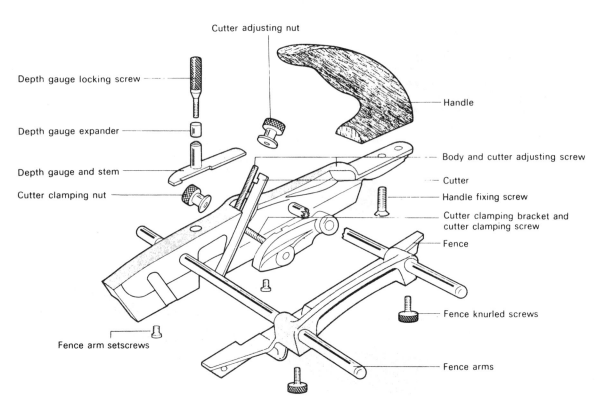

Plough plane exploded to show all parts

38 mm wide and has no cap iron. It goes through the full width of the sole (which means that the plane is open mouthed) and is secured by a screw cap. The length of the plane is 255 mm. In some models adjustment of cut is by a lever, while on others it is by turn-screw. Its use is limited to forming rebates. For this purpose the plane is fitted with an adjustable fence for accurate position of cut and a depth stop which ensures that correct depth is achieved. The fence comprises a piece of metal parallel to the length of the plane which is held against the timber being rebated to ensure that the distance is constant. There are also two cutter positions. The central one being for normal work and the foremost one is to enable the plane to get into tight corners.

Plough planes also have specific uses, but are essential in work carried out by hand. These planes also have adjustable fence and depth stops. Cutter sizes range from 3 mm up to 20 mm. This tool is sometimes useful in forming large rebates by accurately cutting a series of plough grooves and then finishing with a rebate plane.

Chisels

Chisels are an essential part of the woodworker's tool kit. The type and size depends on the work being done and here again it is good sense to use only that chisel which is intended for the purpose. Most types are available in a full range of sizes, from as small as 4 mm to as large as 38 mm. A mallet should always be used when striking a chisel, whatever the type of handle.

Firmer chisels are sturdy, general use tools, which can have either wooden or durable plastic handles. They are made for heavy duty work and if kept in good condition will give excellent service.

Firmer chisel — general use

Bevel edge chisels are obtainable in the same range of sizes as firmers, but are essentially for bench work where accuracy and high quality are demanded. The blade is bevelled on each side to enable the tool to perform fine, intricate work with precision.

Bevel edge chisel — bench work

Paring chisels are bevelled edge chisels, but with a much longer blade of up to 300 mm. This enables the chisel to be used in situations where others are unable to reach.

Paring chisel — accurate work

Mortise chisels are produced specifically to chop mortises by hand and are therefore made from very stout steel. Even so, it would be foolish to think that they can be used without fear of breakage. A good operative will keep the use of even such a sturdy tool as this within its capabilities.

Mortise chisel — heavy duty

Gouges are rather like curved chisels. They can be obtained in various sizes.

Firmer gouges have the cutting bevel on the outside of the blade and are used for finishing curves of external form.

Firmer gouge — outer curves

Scribing gouges are used for internal curves and have the cutting bevel on the inside. This type of gouge can only be sharpened with a slip stone which has rounded edges enabling the whole of the cutting edge to be sharpened efficiently.

Scribing gouge — inner curves

Screwdrivers

Screwdrivers play an important part in the day to day work of carpenters and joiners. As a number of patterns are available it is essential to emphasise the need to use *the right tool for the job in hand.*

Plain screwdrivers are made in various sizes to suit different sizes of screw. The handle may be of hardwood or unbreakable plastic.

Plain cabinet screwdriver

Stubby screwdrivers are used in situations where a long screwdriver cannot be used.

Stubby screwdriver

Ratchet screwdrivers have proved to be popular because they enable the operative to insert the screw with one hand, leaving the other hand free to hold the object in position. Small and large ratchets are available, but the small pattern should not be used on large screws as damage to the ratchet mechanism may occur.

Ratchet screwdriver

Pump screwdrivers are used in an effort to obtain speed of operation and are therefore particularly useful in repetitive operations. They combine a ratchet and pump action and a range of sizes are made to suit most types of work.

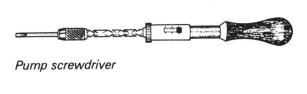

Pump screwdriver

Phillips screwdrivers, or *Posidriv* screwdrivers, are needed for use on those screws which have a star head inset instead of the standard slot head. Various sizes are produced.

Phillips screwdriver

Hammers

Claw hammers should be of the best quality if they are to perform safely and efficiently. The handle (or shaft) should be perfectly fitted to the head. The claw is used to withdraw nails and pins and for this reason must have a taper which is regular and goes to a fine point. Head weights of 16 to 24 ozs are obtainable.

Warrington hammers are sometimes simply called 'joiner's hammers'. They are a cross-pein hammer of smaller size and weight and are used for lighter work, usually on the bench. Pein is the name given to the end of the hammer head opposite the striking face, they are sometimes rounded for special purposes. The cross-pein assists in starting a small pin or brad. Head weights range from 6 to 16 ozs.

Mallets are produced primarily for use with chisels, but are also used for other purposes where the material to be struck is other than metal. Because of its fine, strong texture beech is the most popular material from which mallets are made. To extend their serviceable life an occasional soaking in linseed oil will have a beneficial effect.

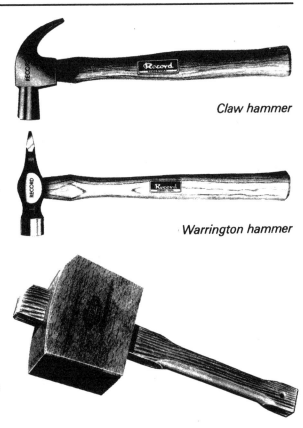

Claw hammer

Warrington hammer

Mallet with beech head

Hole boring tools

There are countless reasons why holes may be required in timber during the course of the carpenter and joiner's work. It may be to enable a screw to pass through from one piece to another or it may be to clear waste material when chopping through a mortise. Whatever the operation, there are tools designed to carry out the work accurately and efficiently.

Ratchet brace

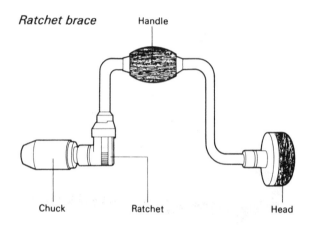

The ratchet brace

This type of brace is now used by the majority of woodworkers. The most popular is the 255 mm sweep, but smaller and larger sweeps are obtainable. Good quality braces have ball bearing heads and three-way ratchets. Universal jaws are obtainable which will receive both tapered and straight shank bits. The ratchet enables the operative to use the brace for insertion or withdrawal in positions where the full sweep cannot be used.

Bits

The bits used in conjunction with these braces vary in length according to their diameter and type. Some have a specific function to carry out.

Twist bits are the most common. There are two patterns:

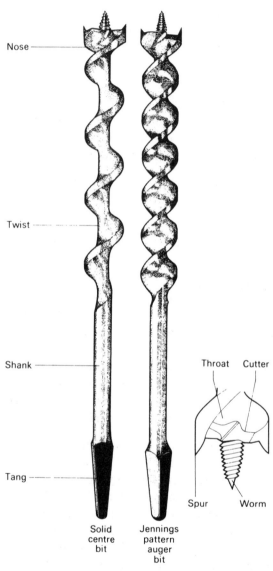

Solid centre and Jennings twist bits

1   *Jennings pattern* is for very accurate work. This bit has a double spiral twist and is available in sizes from 5 mm up to 38 mm.
2   *Solid centre pattern* is for general work. It has only one spiral twist and ranges in size from 5 mm up to 25 mm.

Both patterns have a central threaded worm which draws the bit into the wood, while the spur on the cutters scribe the cut which outlines the hole.

Centre bits provide an excellent means of boring accurate, shallow holes in both hard or soft woods. Both old and new patterns are still available, in sizes ranging from 5 mm up to 50 mm. The new pattern has a threaded worm at the cutting head similar to the twist bit, but with only a single spur.

Forstner bits are designed for fine, accurate work where a flat bottom is required. This bit has no threaded worm and is guided by its outer cutters. It is of great value for any recess work or pattern making. The bits are available in 10 mm to 50 mm sizes.

Expansive bits can bore holes of large diameter in softwood only. Cutters are set by a simple screw setting. They are made in two sizes:

1   Small — for holes up to 38 mm diameter
2   Large — for holes up to 75 mm diameter

Countersink bits are primarily used to provide a sinking to enable a screw head to pull in flush with the surface. Sizes range from 10 mm up to 25 mm. Two patterns are available: *rosehead* and *snailhorn*.

Wheel braces with morse drills are used to make the large number of very small holes required to be drilled by carpenters and joiners. Drills

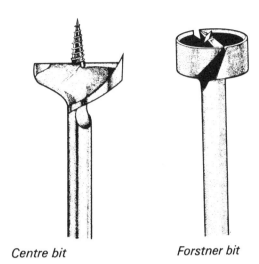

Centre bit   Forstner bit

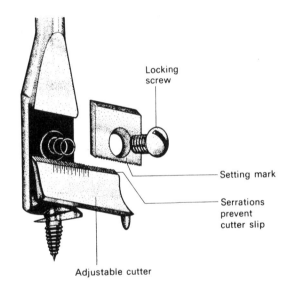

Locking screw

Setting mark

Serrations prevent cutter slip

Adjustable cutter

Expansive bit

Double pinion wheel brace or hand drill

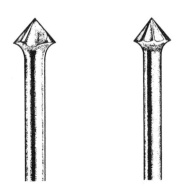

Snailhorn (left) and rosehead countersink bits

can be obtained in many sizes and patterns, including rosehead countersink drills. It is by this means that the vast majority of screwing operations are carried out. The chuck capacity of these braces is only 7 mm in some cases and this obviously restricts the amount of operations they are capable of performing. Double pinion wheel braces give more accurate and longer service than single pinion tools. Some manufacturers call this tool a *hand drill*.

Other useful tools

A *bradawl* will very often be large enough to make a hole to receive a screw, as well as sometimes being used as a screwdriver on very small screws. A small and large bradawl will be sufficient for most operations.

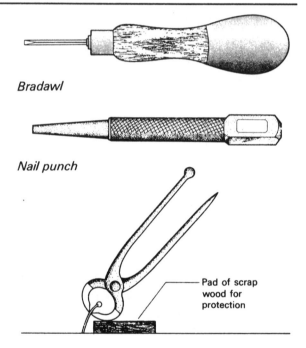

Bradawl

Nail punch

Nail punches are an asset for use on fine pins as well as large nails. The job is never completed until the pins or nails have been punched-in below the surface.

Pincers are necessary for the removal of pins, nails, screws etc. Damage to the workpiece can be prevented by placing a pad of waste wood underneath the pincer head before pulling the nail.

Pad of scrap wood for protection

Pincers in use

Oilstone
Frequent sharpening of planes and chisels will mean that a good quality oilstone is essential. Carborundum stones with a fine cut on one side and coarse on the other are very suitable. For safety the stone should be housed in a hardwood box with a lid. Always use the whole of the stone's surface when sharpening to avoid unequal wear on the stone which will lead to unsatisfactory sharpening in the future.

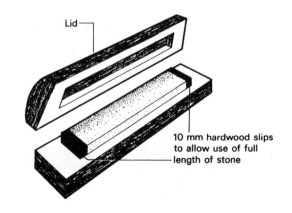

Lid

10 mm hardwood slips to allow use of full length of stone

Spirit levels
Spirit levels also prove useful, especially on site work. There are various sizes and patterns available, but most combine the plumb (vertical) vial as well as the level (horizontal) one. Many also have adjustment on the vial to enable them to be kept accurate by occasional

Oilstone in hardwood case

testing. Each vial comprises a sealed glass tube, normally 6 to 10 mm in diameter, containing spirit. A small bubble in the spirit indicates the required position.

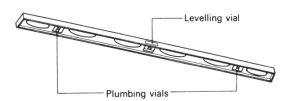

Lightweight spirit level

Cabinet scrapers

Cabinet scrapers are especially useful in the finishing of hardwood joinery which is to be french polished. They also provide the only way to deal with some areas of difficult grain or knots. They should be of the finest quality steel. Sizes range from 100 × 70 mm to 150 × 75 mm.

Sash and 'T' bar cramps

Used in the final assembly of joinery work, these cramps provide the pressure required to hold joints together until they are secured by either wedges or the setting of glue.

Sash cramps consist of a straight mild-steel bar with screw head at one end. The slide thrust can be positioned for width as required by means of a pin fitted into holes in the bar.

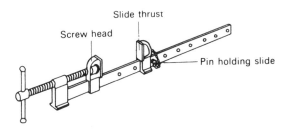

Sash cramp

'T' bar cramps are similar to sash cramps and have similar uses, but because the bar is moulded in a 'T' shape they are capable of heavy duty work and will not deflect when pressure is applied to the screw head. These cramps are available in a range of lengths up to 2135 mm (7′0″), but they can be extended with lengthening bars up to an additional capacity of 1525 mm (5′0″).

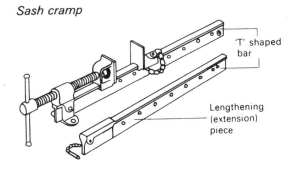

'T' bar cramp

'G' cramps

These are used for much smaller operations. The name is derived from the outline of the cramp which has a curved frame and straight screw head. Its many uses include holding small work to the bench, leaving the hands free to use the tools, and for holding joints in small work together until they are held by glue or other means. 'G' cramps are available in a variety of sizes to suit most situations.

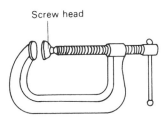

'G' cramp

Questions

1 List the five most commonly used saws, describing their features and uses.

2 Which type of plane is most suitable for finishing work and why should the cap-iron be finely set?

3 Which chisels should be used when working on good quality hardwood joinery?

4 List the advantages of using a marking knife.

5 Describe the method used to check the accuracy of a try square.

6 Describe the different uses of:
(a) a marking gauge,
(b) a mortise gauge.

7 Describe the difference between a firmer gouge and a scribing gouge.

8 Why are hardwood slips inserted in an oilstone box at each end of the stone?

9 Describe the main features of a rebate plane and state their function.

10 Describe the characteristics and uses of:
(a) a firmer chisel,
(b) a mortise chisel.

3 Portable powered hand tools

Portable powered tools have established a permanent place in the range of tools used by woodworkers, both on site and in the shop. For this reason the operative should adopt a responsible attitude to their use. By the correct use of these tools it is possible to carry out many jobs in quicker time and in other cases better quality can be achieved, especially in some difficult or awkward situations. The one thing that must remain uppermost in the woodworker's mind when using an electric tool is *safety*. Electric powered tools bearing the *British Standard Kite Mark*, accompanied by the double square symbol, are double insulated and therefore much safer than tools without this protection. The power supply required (e.g. 240V or 110V) will always be clearly marked on the tool and must be strictly followed. Where the only available supply is 240V and the tool is designed for 110V a transformer must be used to lower the voltage.

Serious accidents can occur with power tools. It is wise, therefore, to adhere to a code of safety which will minimise mishap or accident. At no time should tools be used in such a condition, or in such a way, that danger to the operator is increased. The following points serve as a sound base in an attempt to gain maximum benefits

from the use of portable powered tools without accident.

1 Only the supply voltage which appears on the nameplate of the tool should be used.
2 All three-core cables must be properly earthed.
3 Leads must be correctly fitted to both the tool and the plug.
4 Cables should be kept clear of the cutting head of the tool and generally looked after to prevent damage.
5 Make sure that the plug is properly inserted into the socket.
6 The plug must be withdrawn from the socket when making any adjustments.
7 Do not tamper with electric machines.
8 All known defects should be reported and remedied before use.
9 Never remove a guard to use the machine.
10 Goggles and other protective clothing should be worn.

In addition to these points, no person should attempt to use a portable powered tool without having had some instruction in its use.

Drills

Palm grip drill

This is probably the most widely used of the portable electric tools. They are compactly made and light in weight. Their versatility enables them to carry out a multitude of tasks in timber as well as steel, alloy and masonry.

The design enables the user to apply pressure with the palm of the hand behind the drill bit and so assist in the cutting action. Continuous pressure can be dangerous, however, so the drill should be eased off at intervals to allow clearance of dust from the hole.

These drills are available with single or two

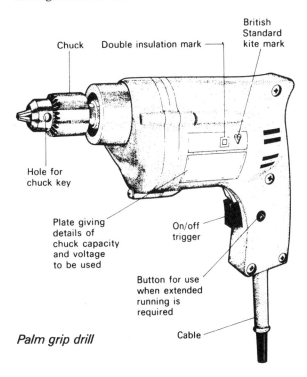

Chuck   Double insulation mark

British Standard kite mark

Hole for chuck key

Plate giving details of chuck capacity and voltage to be used

On/off trigger

Button for use when extended running is required

Cable

Palm grip drill

speeds. The two-speed models have a slower speed which is particularly useful when using a tipped drill for boring into brickwork or masonry. Drill bits should be fully inserted into the chuck and tightened with a chuck key of appropriate size for the tool being used. The chuck has three holes for this purpose and it is essential that they are all tightened.

Where the work to be done involves a great deal of repetitive drilling the drill can be fitted into a *drill stand* which will ensure a much greater rate of production, without loss of accuracy. The *Health and Safety at Work Act* 1974 (HSAWA) insists that a chuck guard *must* be used when the drill is attached to a drill stand.

Correct method of holding drill

Back handle drills

These drills carry out much the same operations as the smaller palm grip drills, but are capable of larger capacity drilling, hole forming etc. They are, in effect, a heavy duty version of the palm grip drill. These tools can also be mounted into a drill stand if required. They have a larger chuck capacity and can therefore bore holes of greater diameter when required. The tool is capable of sustaining heavy pressure for longer periods but, again, clearing of dust at intervals is recommended. Back handle drills are obtainable in single, two and four-speed models to suit virtually any type of work. A handle is attached to give greater control during use.

Rotary percussion (hammer) drills

Hammer drills are manufactured in the palm grip or back handle pattern. In addition to normal drilling, the tool can also provide a percussive hammer effect on the bit to enable it to overcome obstinate or hard materials. This percussive action is optional and is brought into action by means of a switch clearly marked on the tool. This capacity for heavy work makes the tool ideal for drilling into masonry or concrete when fixings are required.

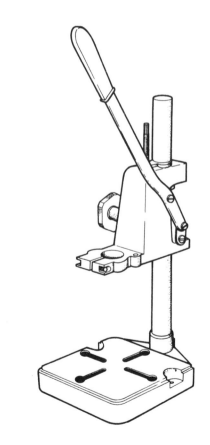

Bench stand drill

Rotary percussion drill — using percussive action to drill into reinforced concrete

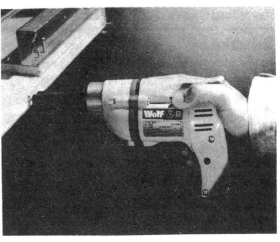

Palm grip screwdriver

One model is obtainable with variable speed and 10 mm chuck capacity while others have two speeds and chuck capacity up to 16 mm.

For work where the depth of hole is of some importance a side handle with depth gauge attached is available and can be used on most models.

Screwdrivers

Screwdrivers for repetitive production processes are available in palm grip and back handle patterns. Fasteners are made to suit standard or star headed screws. These tools are equipped with a depth setting device which ensures that the tool will stop driving when the fastener has reached the correct depth. All models have a reversing switch for withdrawal of screws and the back handle model has two speeds to meet most requirements.

Something which would prove useful in the very small workshop is the palm grip drill mounted in a *mortise stand*, with accessories made to cut mortises up to 12 mm and wider if cuts are repeated. Only the mortise chisels made for this purpose should be used.

Mortise stand

Powered saws and planes

Portable saws

Portable saws are also used extensively to perform a variety of tasks. These include ripping, cross cutting and even cutting so finely as to dispose of the need to plane or clean up. However, the saw must be fitted with the correct type of blade for each operation. In addition, abrasive discs can be fitted to cut stone, brick or steel if required. These saws are fitted with a soleplate which is adjustable to tilt at between 0 – 45°. They have an easy grip handle and grip knob to give the operator maximum control at all times.

The blade is shrouded with a wrap-around automatic telescopic guard which must be functioning efficiently at all times to ensure that the revolving blade is never fully exposed. An adjustable side fence can also be used for safe and accurate parallel cuts. The depth of cut can be pre-determined by a simple lever adjustment. Two hands must be used on this tool at all times. The material being cut should be secured in such a position as to give maximum control to the operator to minimise the possibility of accidents.

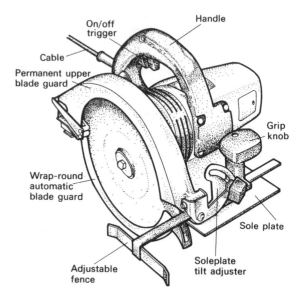

Portable saw

Portable saw in use

Portable saw being used for angular cutting — note that cable is clear of saw

Provided that the correct blade is fitted the portable saw will perform efficiently for any length of time. The maximum thickness of timber that can be cut with a 235 diameter blade is 84 mm. However, if the soleplate is in the maximum tilted position of 45° the maximum thickness of cut is reduced to 62 mm.

Blade diameters are available up to 235 mm and, as with hand saws, the teeth size and set will govern the fineness of the cut. Tungsten carbide tipped saw blades can be obtained which will give much longer service before sharpening is necessary.

Dust control equipment is also produced for some models of portable saw. This provides an excellent safeguard when cutting materials which may contain harmful dust, especially asbestos. This equipment satisfies the demand for strict safety limits imposed by the *Asbestos Regulations* 1969, provided that it is connected to a suitable industrial suction unit.

Dust control shroud

Jig saws

Jig saws are very popular for cutting shapes etc. and can be efficient and accurate on hard or soft woods up to 60 mm in thickness. Better models have two speeds, which increases the tool's ability to cut a great many materials with the same efficiency. The soleplate has a secondary position which enables the saw to be used for getting in close to return surfaces. A wide range of blades for various materials makes the jig saw very versatile. Once again, the variation of blade is in the number of teeth per 25 mm. Blades with up to 32 teeth per 25 mm can be obtained.

Although the saw is operated by one hand only, the workpiece should be firmly fixed while it is being cut. The saw will carry out virtually any shaped work in hard or soft woods up to a maximum 60 mm thickness. Provided it is not overworked it will continue to cut fast and

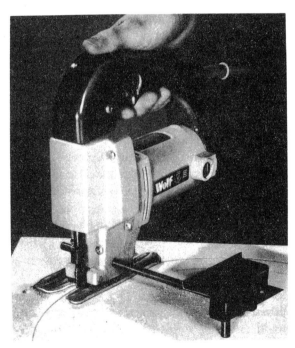

Jig saw in use with jig being used for accurate radius cutting

accurately for a considerable time. Full circular work is made easier with the use of a cutting guide which is supplied with some tools as standard equipment. To enable accuracy to be maintained some jig saws are fitted with an air flow directed at the point of cut in order to blow away the sawdust as it builds up. This addition is very useful and increases the safety to the operator by making it unnecessary to remove the sawdust by other methods.

Powered plane

The powered plane can be of immense value when a large amount of planing is required on site and is needed quickly. The planing action is not dissimilar from that of the ordinary jack plane, except that it has a high speed twin-blade cutting block capable of removing waste very rapidly and at the same time leaving a very good finish on the work. Surface planing is, therefore, carried out quite simply. In addition, the plane will work chamfers and rebates as desired. To ensure accuracy, the plane is equipped with a long soleplate which also affords greater control when forming a rebate or chamfer. The front position of the soleplate is adjustable and the depth of cut, up to a maximum of 3 mm, can be set by the use of a control knob on the top side. This knob also functions as a hand grip when the plane is in use. Depths greater than 3 mm, such as for rebating etc., can be accurately achieved by setting the depth gauge on the side of the tool to the required position. An adjustable fence can also be used for rebating, as well as for additional control when planing narrow edges.

Planerette

Planerette is the name given to a smaller plane which has many of the same characteristics and uses as the larger powered plane. Its size does, of course, make it easier to accommodate in the tool box and it is also of lightweight design. This makes it possible to use the plane in positions in which it would be unwise and unsafe to use the larger tool.

The planerette has a 430 watt powered motor driving a nylon vee belt, which means that no oil is required as part of its maintenance. It is essential, however, that the cutters are kept sharp and in good condition.

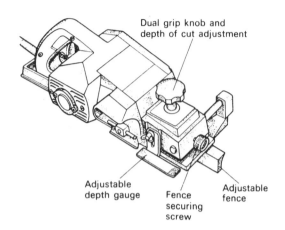

Dual grip knob and depth of cut adjustment

Adjustable depth gauge

Fence securing screw

Adjustable fence

Powered plane

Forming a rebate

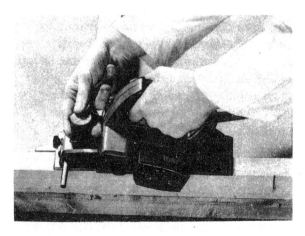

Planerette

Portable routers

The use of portable routers is now quite extensive and yet mechanically they are the simplest of all power tools. A central spindle is driven by the motor and because there is no variation of speed on routers, no gear mechanism is needed.

Despite this, routers are capable of a wider range of work than most other tools. They carry out cutting and shaping operations with equal efficiency. In addition, the router produces a very good finish to all its work, provided that the tool is used with care and skill. This excellence of finish is directly attributable to the speed of the tool, which ranges from 18 000 rpm at the lower end to 35 000 rpm at the upper end. Therefore, it is many times faster than other power driven hand tools. Such very high speed demands greater control and concentration from the operator if accuracy and safety are to be maintained.

The router is an extremely versatile tool. Some of its more important functions are listed below:

1 Beading
2 Laminate trimming
3 Dovetailing
4 Mortising
5 Rebating

The two types of router in common use are:

1 The heavy duty router
2 The heavy duty plunge router

Heavy duty router

The heavy duty router has a set base which is designed to give maximum view of the work being carried out. The depth adjustment is found just above the body of the tool and is in the form of a threaded ring. The motor housing is also threaded to receive the adjustment ring in such a way that it is possible to achieve very

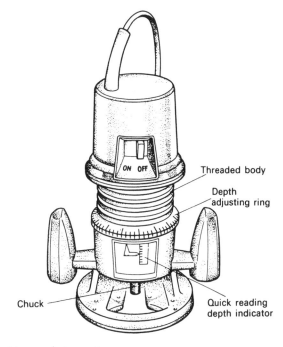

Heavy duty router

fine settings where needed. When the required setting has been made a lever can be used to ensure that no further movement takes place.

The most important difference between this router and the plunge router is the manner in which it has to be fed onto the workpiece. With the cutter projecting from the base, care must be taken to feed it gradually and accurately with a very firm grip to prevent it from snatching. Some heavy duty routers are fitted with a moulded sub-base to prevent damage to the surface of the timber. The 3-position switch automatically locks the shaft for safety when adjustment or changing of cutters become necessary.

Plunge router

By contrast, the plunge router is fed into the workpiece at 90° to the base and withdrawn in the same way. This action is made possible by the two spring loaded plungers fitted to each side of the tool.

As with any other router the depth setting must be carried out prior to use, but three changes in the depth of cut can be set on the turret stop to give instant changes in depth while the tool is in use. The base is designed to give a maximum view of the work and facilitate accuracy. Skilful router work requires more practice and care than most other power tools. Routers are used free-hand or with a templet. Good freehand routing will only come from experience of using the tool.

A very large range of router cutters are now made to perform almost any shape and size of cut. Basically, the two types used are those which are for free-hand or templet work and those having a *pilot* on the tip which acts as a guide when edge work is being carried out. The cutters are made from special steel to give longer cutting life and those made by the best manufacturers will fit any make of router. Apart from cutting rebates and mouldings the router is very useful for trenching or cutting housing staircase strings i.e. the wide boards on each side of the stairs into which the steps are housed and wedged. This type of work should always be done with the aid of a templet to achieve greater speed and accuracy. All forms of trenching and sinkings in timber can be performed by this tool. Dovetailing with a purpose made dovetail templet is also possible.

Work should be properly secured before attempting any operation with the router. The great speed of the router demands that both hands are required to control the tool and are therefore not available to steady the work-piece as well. All-metal tables can be obtained into

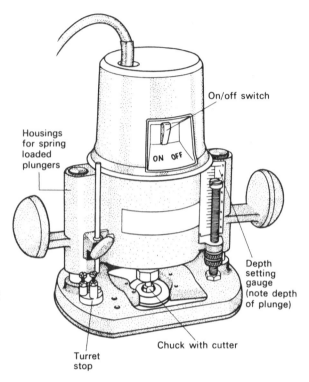

Heavy duty plunge base router

The router being used for stair string housing with the aid of a templet

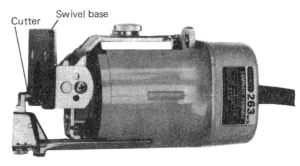

Cutter · Swivel base

Laminate trimmer

which the inverted router can be mounted. This gives the appearance of a vertical spindle, but it must be borne in mind that the capacity of the router remains the same and should not be overloaded. However, some operations are made easier with the table, especially on small repetitive work.

Laminate trimmers

Laminate trimmers are very similar in many respects to routers. The increase in amount of plastic laminate finishes used by woodworkers has meant that these trimmers are also becoming more popular. They have a very high speed of 30 000 rpm and are made in such a way as to make them easy to hold and use provided care is exercised.

These tools can remove an overhang of laminate with ease and accuracy on either blockboard or chipboard. They can also be used to trim and scribe the back edge of a worktop or counter by resting the trimmer against the wall with the material in position. The result is a very good fit against the back edge and this is particularly useful when fitting worktops etc. on site.

A swivel base enables the tool to cut angles of any inclination up to 45°. The laminate trimmer also has a roller guide and fine depth adjustment which can ensure smooth trimming.

The trimmer in use

Sanders

Belt sanders

Belt sanders provide a speedy means of finishing surfaces where a plane would be too large. They incorporate a continuous belt of abrasive paper which rotates about two rollers. One roller is fixed while the other is adjustable. It is this adjustable roller which tensions the belt. Belts can be fine or coarse grit to meet most requirements. Sanding with these machines should be carried out with the grain and not across it. Ridges will form if pressure is applied too heavily or unevenly. Where there is a definite difference in the thicknesses of material this should first be removed by plane before sanding is commenced.

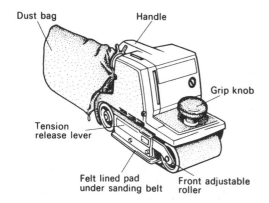

Belt sander

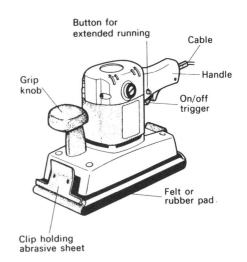

Orbital sander

This type of sander is capable of fairly heavy duty work and is therefore fitted with a dust bag to gather the waste material as it is removed during the operation.

To exert proper control at all times both hands should be used. A front grip knob is conveniently placed to enable the operator to achieve best results and allow pressure to be applied in the place where it is needed. Although various sizes of belt sander are obtainable a model with a 100 mm wide belt is usually suitable for most jobs. A felt lined pad is situated just behind the belt where it comes into contact with the work. Care should be taken to keep this pad in good condition otherwise poor quality finishes will result. Belt sanders are quite powerful, which makes it necessary to keep the work firmly anchored throughout the sanding operation.

Orbital sanders

Orbital sanders are for finishing only and should not be used to remove waste material. They are generally much lighter in construction than belt sanders, but are capable of producing a good quality finish provided skill is used in the process. The base of the tool is fitted with a felt pad which can be substituted for rubber where this is more desirable. Stretched across this pad and retained at each end by a clip

The orbital sander in use

arrangement is the sheet of abrasive. When the machine is switched on the base moves in 3 mm diameter orbits at a rate of 12 000 per minute. Best results will occur if two hands are applied to the tool, but it is wrong to apply too much pressure as this can cause inaccuracy. When the tool appears not to be cutting sufficiently well a check should be made to see whether a coarser grit sheet should be fitted.

It is advisable when using any portable powered sanding tool to use the belts or sheets which are manufactured specifically for the tool. Accidents can and do occur when makeshift belts or sheets are fitted.

The Cordless Range

In addition to the mains-powered tools seen in the previous section there is a range of battery operated (cordless) tools which the carpenter and joiner will find extremely convenient and useful, particularly in situations where no mains power supply is available. There is no trailing cable to cause problems when working some distance from the power point or from ladders etc. at a high level.

The range consists of drills, screwdrivers and sanders which can have wider applications when the appropriate accessories recommended by the manufacturers are used. Battery power, like electricity, is expressed in volts (V), e.g. 10V, 12V etc., indicating that the higher voltage tools are more suited to heavy-duty work. Batteries are rechargeable using the charger supplied when purchased. Although recharging takes only a few hours, depending on the model, it is a sensible policy to have one battery on charge while another is in use if the tool is in continuous use.

Selection of the appropriate tool for the work is important because accidents and damage can occur when forcing a tool designed for light work to carry out heavy-duty operations.

Drills/Screwdrivers

These resemble their mains-powered counterparts but have battery attachments. Other features include:

1 Chucks can be of the traditional type with a key or self locking (keyless) variety.
2 Two speeds, fast or slow to suit the material.
3 Variable speed, so the operator can start slowly and increase speed as required for accuracy.
4 Reverse function, very useful for removing screws or clearing holes while drilling.
5 Torque control, with a wide range of settings to enable every need to be met accurately.
6 Impact (hammer), which is very useful when drilling into obstinate materials.

Maximum benefit from the use of these tools will only be achieved by using them sensibly and safely. This will mean the appropriate use of safety clothing, e.g. goggles, gloves etc. as well as the correct drills and screwdriver bits.

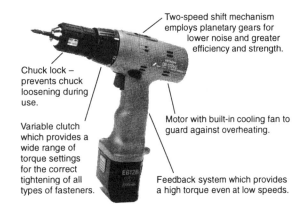

Cordless drill/screwdriver, showing typical features

Questions

1 List as many points of the safety code as possible, outlining the dangers that exist if they are ignored.

2 What do the *kite mark* and *double square symbol* indicate?

3 Why is it better to drill holes in short bursts rather than continuously?

4 To what type of work is a percussion drill best suited?

5 Why is it necessary to secure material being cut with a portable saw?

6 What is the advantage of using a two-speed drill?

7 What is the function of the wrap-around guard on a portable saw?

4 Basic woodworking joints

The basic joints used in woodwork are required to perform three main functions:

1 *Framing*, such as that required for panelled doors etc.
2 *Widening*, where timber widths are insufficient and two or more boards are joined together.
3 *Lengthening*, to increase the length of a beam etc.

It is most important to observe the basic rules for setting-out these joints because their strength depends upon it. Any variation to suit particular requirements must always ensure that the efficiency of the joint is in no way diminished.

Although machinery is used widely in joint construction this section is concerned with showing how the woodworker will construct them by hand operations.

Framing joints

Halving joints

Halving joints are probably the simplest of the framing joints. They are used to form external angles, called *corner halving*, and also for intermediate members, when they are termed *tee halving* joints. The name of the joints is taken from the fact that the thickness of the material to be used is halved (half being removed and half remaining), but the width of the material is normally left in full. This type of joint is not self supporting and is therefore best used where it is to be clad with a sheet material on one or both sides.

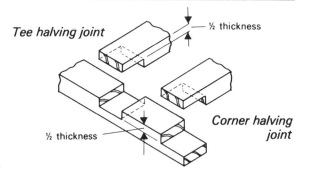

Tee halving joint

½ thickness

½ thickness

Corner halving joint

Housing joints

Housing joints are extensively used in many forms of framing. The important thing to remember with these joints is that the proportion of the thickness used is one-third. These joints can also be used for outer framework as well as intermediate joints.

If an angle joint is required a *tongued housing* will be found suitable. The tongue being that part which projects into a housing of similar size. The tongue projects one third of the thickness of the piece into which it is housed, while its own thickness is one-third of the material being used.

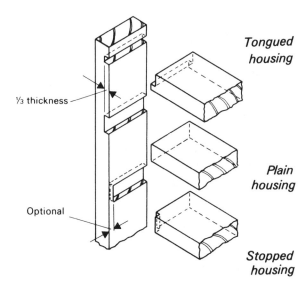

⅓ thickness

Optional

Tongued housing

Plain housing

Stopped housing

Plain housing joints are used for intermediate rails. They are formed by leaving one member full in width and thickness while the housing is one third of the thickness of the piece into which it is framed.

Stopped housings are only a slight variation on a plain housing. The housing is stopped back from the face by an optional amount to give the appearance of a clean joint on the face of the work. The optional distance only needs to be a small amount to obtain the desired effect.

Mortise and tenon joints
The mortise and tenon joint, with its many variations, is probably the most common and certainly one of the strongest framing joints used. The *plain* or *through mortise and tenon* is the basis for all such joints. It consists of a tongue on one member which is called the *tenon* and a mouth to receive it in the other member called the *mortise*. The width of the tenon is

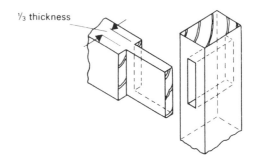

Plain or through mortise and tenon

usually the full width of the material being used, while the *thickness* of the tenon is always *one third* of the material thickness.

The timber is marked out with the use of a mortise gauge and marking knife. If these markings are followed closely an accurate joint should result. Work on the joint is commenced by chopping the mortise through. This will be

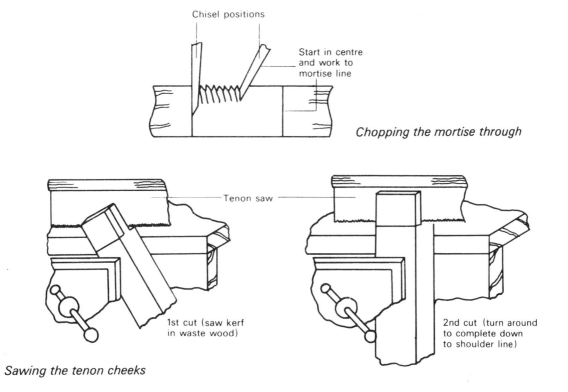

Chopping the mortise through

Sawing the tenon cheeks

easier if it is started in the centre of the mortise and chopped back towards the setting-out mark. Only clear half the mortise depth to begin with, then turn the piece over and repeat the process. The mortise is formed first so that any slight inaccuracy that occurs can be put right by making the necessary allowance when cutting the tenon. The timber is then placed in a bench vice at a convenient angle to saw down the side of the tenon. This should be done on one side first, then the piece is turned round in the vice to cut down the opposite side. Never cut the shoulders before cutting the sides of a tenon.

Stub tenons are used where it is not possible or desirable to carry the joint right through. The depth of the tenon need not exceed one half of the width of the material into which it is going because this makes the work harder and no further benefit will be obtained anyway.

Haunched mortise and tenons enable a very strong joint to be formed at the angle of any frame, such as the top corner of a door. In this joint part of the tenon is cut away, leaving only a haunch in the upper part. The function of the haunch is to prevent the twisting of the rail after the joint is assembled and to give added strength to the joint as a whole. For best results the haunch depth should equal the thickness of the tenon up to a maximum of 15 mm, beyond which it may begin to detract from the strength of the joint. The tenon width should not be less than half the width of the material being used.

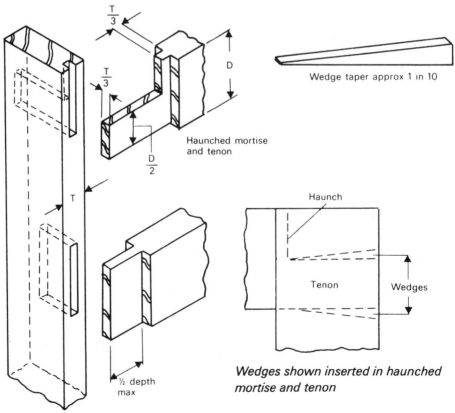

Wedge taper approx 1 in 10

Haunched mortise and tenon

Haunch

Tenon

Wedges

Wedges shown inserted in haunched mortise and tenon

½ depth max

Stub tenon

In many cases *wedging* is an integral part of the mortise and tenon joint. They are used in the great majority of cases where the tenon goes right through. The wedges should not be too sharply tapered, a taper of 1 in 10 is suggested. It is useful to cut these wedges from the waste material in a haunched tenon, since they are already the correct length and thickness. At no time should wedges be cut to a fine point. This could act as a chisel and cause damage when being driven in.

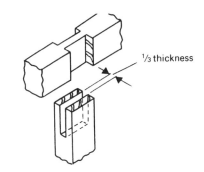

Bridle joint

Bridle joints

Bridle joints have the same proportions as the mortise and tenon and are rather like the mortise and tenon in reverse. The central one-third is left in on one member while the outer two-thirds are cut away. On the piece jointed to it the centre is removed to allow it to fork over and complete a good useful joint to suit many circumstances.

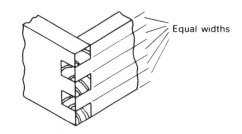

Combed joint

Combed joints

The combed joint is used to secure the angles of boxes, drawers and other carcassing work. It is usually a machine made joint. The width of material being used is divided into an odd number of equal parts. The fingers on the comb need to be accurately cut to get maximum strength from the joint.

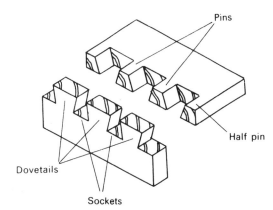

Plain dovetails

Dovetail joints

Dovetails are also used in a wide variety of ways, but they always provide a satisfactory result when they are correctly set-out.

Plain dovetails are used in the manufacture of drawers, cases and boxes of all description. The angle for dovetails in softwood is 1 in 6 and 1 in 8 for hardwood. A fully assembled dovetail joint consists of pins, half pins and sockets. The number of dovetails used on any job is discretionary. It will depend on the width of the board and other factors, but the correct setting-

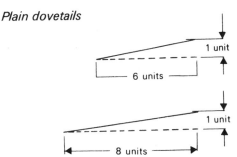

Softwood (top) and hardwood (bottom) dovetail angles

out will always commence and finish with a half
pin. In setting-out, the dovetails must be
marked and cut first and from this the pins are
marked and cut. This stage of the marking-out
is important and care should be taken to ensure
that pins are only fitted into the sockets from
which they were marked. Accuracy must be
maintained throughout, otherwise an unsightly
and weak joint will result.

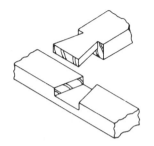

Dovetailed halving joint

A *dovetailed halving* joint is one of the many
variations frequently used. It has much to
commend it when used as a tie rail because the
dovetailed halving will resist the tension placed
upon it.

Lapped dovetails become necessary when the
strength of a dovetailed joint is required, but it
is desirable that they should remain unseen.
The best example of this is on drawer fronts
where the pins can be seen at the sides, but the
face of the drawer is free from any signs of
construction. The amount of lap should not be
smaller than 3 mm for safety, as the danger of
breaking out during construction will be
increased. A lap of 5 to 6 mm will be quite
adequate for most situations.

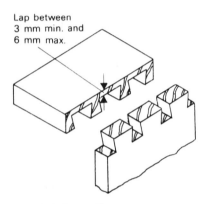

Lap between
3 mm min. and
6 mm max.

Lapped dovetails

Widening joints

The difficulty in obtaining wide boards of good
quality has led to an increase in the number of
times it is found necessary to join narrower
boards edge to edge in order to produce boards
of a required width. Several methods of jointing
these boards are in common use. The method
used will be at the discretion of the woodworker,
but will also depend very largely upon the use
to which the boards will be put.

Whatever method is chosen the edges of the
boards will have to be carefully prepared before
gluing. This preparation is best done with a

jointer plane, as a true straight edge is necessary
if a good quality joint is required.

Rubbed joint

If the jointed board is not to be subjected to
pressure, or if it will be in a position where
supplementary support will be used, as with
shelving, then a simple rubbed joint will be
found adequate. In this method the edges of
both boards must be shot straight, square and
true. The glue is then applied to both surfaces
before they are brought together. When the two

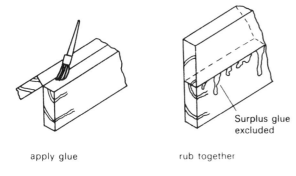

apply glue   rub together

Surplus glue
excluded

Rubbed joint

edges are brought together they should be
rubbed or moved backwards and forwards a few
times to exclude surplus glue. This rubbing
action also creates a vacuum which gives the
joint immediate 'grab'. Light pressure from
sash cramps may be necessary until the glue has
set sufficiently to hold of its own accord.

Loose tongue joints

Loose tongue joints provide additional strength
where needed. Again, the boards must be 'shot'
together as for a rubbed joint. A groove is then
formed in each edge to receive the tongue,
which can be of plywood, timber cut cross-grain
or normal long grain timber. The thickness of
the tongue should never be more than one third
of the boards' thickness and its width will be
adequate if it is four times greater than its
thickness. Plywood tongues are the strongest
and easiest to prepare. This type of joint should
be left in a cramp for about four hours to allow
the glue to harden.

Slot screwed joints

The slot screwed joint is probably the strongest
method of jointing two or more boards. First,
the edges are planed true. Screws are then
placed at intervals along the edge of one board,
but are only screwed in as far as the shank (the
upper part which has no thread). The distance
between these screws will vary with
requirements, but maximum strength will be

obtained from screws 300 mm apart. In the
corresponding positions on the edge of the
other board, holes are bored to allow the head
of the screws to enter. These holes should be
equal in depth to the projection of the screw
with 2 mm extra for clearance. From the centre
of these holes slots are cut which should be
slightly larger than the diameter of the screw
shank. The slots should be about 18 mm in
length and slightly undercut at the end to allow
for the screw head. The boards are then placed
firmly together and driven along as shown.
This enables the screw heads to bite into the
sides of the slots, thus forming a very sound
joint.

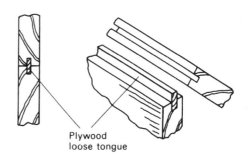

Plywood
loose tongue

Loose tongue joint

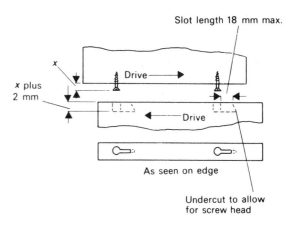

Slot length 18 mm max.

x

x plus
2 mm

Drive

Drive

As seen on edge

Undercut to allow
for screw head

Slot screwed joint

Lengthening joints

Lengthening joints sometimes become
necessary for various reasons. The method used
to join two or more pieces of timber together
end to end will, once again, depend on the
circumstances prevailing. Nowadays, large
sections of timber are made up by gluing many
pieces of timber together, this is called
laminating.

Beam lamination

In laminated beams a simple butt joint will be
sufficient, provided that the butt joints are
staggered sufficiently to give greater 'bonding'
to the layers of timber being used.

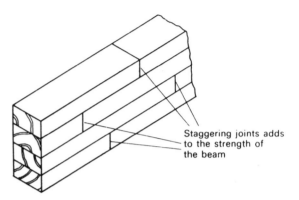

Staggering joints adds
to the strength of
the beam

Beam lamination

Scarf joints

Various scarfing joints have been used to
overcome this need in the past, but in recent
times the one described below has been most
widely used.

The *splayed and wedged scarf joint* produces a
very strong joint if correctly set-out. The total
length of the jointed area should not fall below
three times the depth of the beam and this will
give the angle of splay. The illustration shows
that the hardwood wedges are one sixth of the
total depth of beam and each member is
'squared' off at its end. It is important that this
squaring off is at 90° to the splay, because
when the wedges are fully driven in they will be
forcing each end of the joint into the housing
formed for it. This produces a maximum hold
on the extremity of the joint.

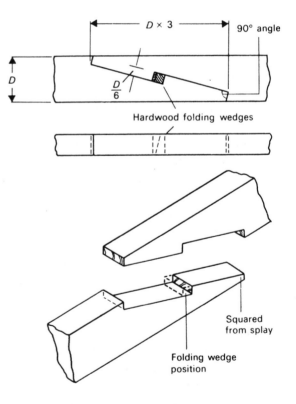

Splayed and wedged scarf joint

Questions

1 Where is a *haunched mortise and tenon* used and what proportions are used in forming the joint.

2 Describe the angle of dovetails suitable for (a) hardwood and (b) softwood, and give reasons.

3 With the aid of sketches describe the principles of a slot screwed joint.

4 Sketch an *angle joint* formed with a tongued housing, showing correct proportions.

5 To a scale of half full size, draw a lapped dovetail joint showing three dovetails using 100 × 20 mm material.

5 Woodcutting machinery

Wherever joinery is manufactured, however small the premises, there is bound to be at least one item of woodcutting machinery. In some establishments there are very many machines, some very large, carrying out so many different operations that only the assembly is left to be completed on the bench. This process is called *mass production* and is common to the manufacture of kitchen units and other furniture.

In this section, however, we shall only consider the type and size of machine which is likely to be found in workshops of small or medium size and output. It is in these types of establishment that the woodworker may be required to set up and use these labour saving machines in the interest of faster production. It must be remembered however, that woodcutting machinery carries a much higher accident risk if not correctly set up. *All appropriate guards and safety devices must be in position before use.*

The Provision and Use of Work Equipment Regulations 1998 (PUWER 98)

The *Woodworking Machine Regulations 1974* have now been totally replaced by the *Provision and Use of Work Equipment Regulations 1998* (PUWER 98) which cover all equipment in the workplace, not just in woodworking establishments. Because of the high risk potential associated with woodworking machines, the Health and Safety Commission (HSC) has drawn up an Approved Code of Practice (referred to as ACOP) which will give valuable guidance to providers (employers) and users (employees) and assist in their compliance with the new regulations. Some of the previous regulations are retained in the new ACOP because they still meet the requirements.

There is provision within the ACOP to allow until the year 2003 for some machinery to be brought into line with the requirements of PUWER 98. This relates to limited cutter

projection on planers and efficient stopping (braking) devices for all machines. The primary purpose of these regulations is to minimise the risks to health and safety of all in the workplace. It is for this reason that Regulation 9 sets out a rigorous training programme which, if followed correctly, will ensure that no-one will attempt to use any woodworking machinery before demonstrating their competence to do so by way of assessment.

Regulation numbers quoted in the following text refer to the appropriate PUWER 98 regulation or ACOP guidance note.

Regular and thorough maintenance of *all* machinery must be carried out. The frequency of use will determine the programme which must ensure the highest degree of safety in the use of woodworking machinery.

Circular Saws

A summary of the ACOP requirements for these machines is listed:

1 That part of the saw blade which is below the table shall be guarded to the greatest extent possible. (**Reg. 11**)

2 A strong, rigid and easily adjustable riving knife must be securely fixed directly behind the line of the blade with a gap of 8 mm max. between the knife and the saw blade at bench level. The function of the riving knife is to separate the timber as it passes through the saw and prevent it from jamming. (**ACOP 12**)

3   On a saw of 600 mm diameter or less, the riving knife shall extend above the saw table to within 25 mm of the highest point of the blade. On a saw larger than 600 mm the riving knife must extend above the table by a minimum of 225 mm. (**ACOP 12**)

4   That part of the saw blade above the table must be fitted with a strong and easily adjustable guard which must be positioned as close to the workpiece as possible. (**ACOP 11**)

5   An adjustable extension piece should be fitted to the front of the guard. The flange of the extension piece must extend below the roots of the teeth. (**ACOP 11**)

6   Ripping operations are to be carried out *only* when the saw blade projects through the upper surface of the material. (**ACOP 4**)

7   A notice must be fixed to all circular saws specifying the diameter of the smallest saw allowed on the machine. (**ACOP 23**)

8   Where the rundown time is more than 10 seconds a suitable stopping (braking) device must be fitted. (**ACOP 15**)

9   A suitable push-stick, 450 mm in length and tapering from 25 mm thickness at the hand-held end to 12 mm thickness at the 'bird's mouth' end, shall be provided and be available for use at all times. (**ACOP 8**)

Circular saws are found in the majority of establishments concerned with the manufacture of woodwork. If there is only on present it will be either:

1   The *heavy duty circular saw bench*, or
2   The smaller *dimension saw bench*

Heavy duty circular saw bench

The heavy duty saw bench, as the name implies, is capable of sawing operations on large sections of timber. However, it cannot achieve the degree of precision possible on the smaller, but more accurate, dimension bench.

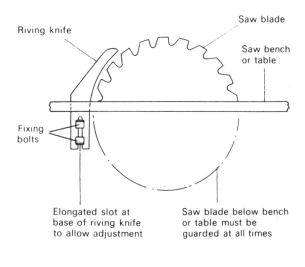

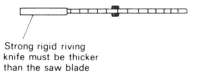

Detail of circular saw bench

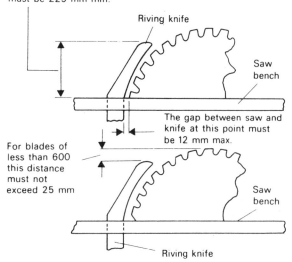

Circular saw regulations

Heavy duty saws are manufactured to take saw blades of up to 800 mm. The teeth on blades intended for ripping are designed differently from those used for cross-cutting. The hook of the tooth is angled backwards.

The hand-feed saw bench has two saw spindle speeds which give the saw a wide working range capacity, these are 1250 rpm for the 800 mm diameter blade and 1520 rpm for the 660 mm diameter blade. The saw spindle itself can be raised or lowered by means of a handwheel which is located in a position where is it easy to operate.

A robust *fence* is incorporated which can be tilted at any angle up to 45°. This makes the cutting of splays or chamfers relatively simple. The fence can be turned over to make the table clear for the cutting of sheet materials.

To comply with ACOP 11 a *crown guard* is fitted. This is also of strong construction. It is also known as the *top guard* and is fitted with an easily adjustable front shield so that the operator is given maximum protection. The regulations require that this guard should be positioned so that it is as close to the workpiece as possible.

The riving knife rises and falls with the saw spindle, but needs to be adjusted when a saw of different diameter is fitted. The riving knife acts as a guard to the teeth on the back of the blade as well as separating the timber. All riving knives must be two gauges thicker than the thickness gauge of the saw blade.

Two very common operations performed on the circular saw bench are:

1 Deeping: this is when the timber is being ripped through its widest dimension.
2 Flatting: the ripping is done through the narrowest part of the timber.

When using the saw bench it must be remembered that at no time should the hands be placed nearer than 300 mm to the saw blade.

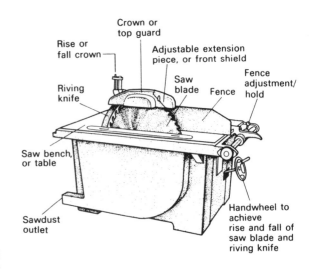

The heavy duty circular saw bench

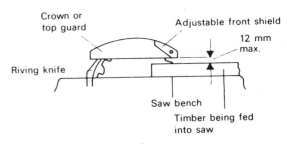

Position of crown guard

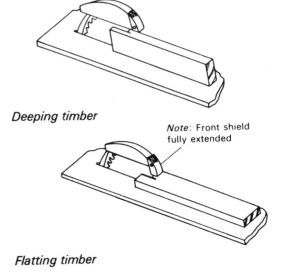

Deeping timber

Flatting timber

Safety aids

To make it possible for circular saws to be used as safely as possible full use should be made of the following aids. For greater reliability they should be made from hardwoods.

1   Push-stick: Approx. 450 mm long, the push-stick has a 'bird's mouth' cut in one end to enable its use without slipping off the material being cut. It must be used to prevent hands getting nearer than 300 mm to the blade and to remove all workpieces less than 150 mm wide from the saw. (**ACOP 8**)

2   Push-block: When cutting short, but deep, sections of timber the push-block enables the operator to keep hands at a safe distance from the saw blade on the off-cut side, while the push-stick is used to feed the material through the saw.

These safety aids should be used at all times and be kept in a convenient position near the machine.

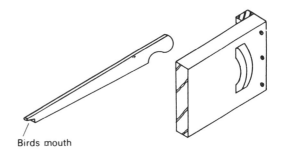

Birds mouth

Push stick   *Push block*

Dimension saw bench

The *dimension and variety saw* is a smaller type circular saw bench capable of carrying out all the work done by the heavy duty machine, but on a smaller scale. In addition, it is designed to cut sheet material to a high degree of accuracy. The scope of the work possible on this machine is further increased by the fact that the saw spindle has vertical adjustment from zero to maximum for depth of cut or trenching head, as well as being tilted up to a maximum 45°. The fence is also adjustable up to 45° and will allow

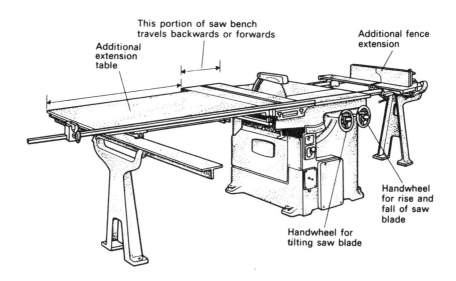

This portion of saw bench travels backwards or forwards

Additional extension table

Additional fence extension

Handwheel for rise and fall of saw blade

Handwheel for tilting saw blade

Dimension saw set up to cut sheet materials

a maximum width of cut of 762 mm.

The cutting of sheet materials is made easier by a table (or bench) which is in two portions. On the right of the saw blade the table is permanently fixed while the portion to the left is made to travel on rollers with an easy action. A simple locking device secures this part of the table when it is not required for use. Both portions have a dovetailed groove which receives a mitre gauge. It is on this gauge that extremely accurate splayed cuts can be carried out: cuts which would require no further fitting. Guards and riving knife on this type of saw are identical to those on the larger saws.

Several types of saw blade can be fitted which will allow various operations to be carried out speedily and efficiently. For instance, if ripping is required so that the sawn pieces have a very clean face, or if plywood is to be clean cut, then a saw blade with a *novelty tooth* should be used. This type of blade is also hollow ground across the plate. If on the other hand it is necessary to achieve a very clear cut, either with or across the grain, different saw blades can be used having teeth which are more suitable for this task. An extension table is also available for the dimension saw bench when large quantities of sheet material are required to be cut very accurately.

The *cross-cutting and trenching machine* combines a number of valuable uses. Simply as a *cross-cut saw* it performs a very useful function when timber is taken from the storage rack and brought to length in order to minimise waste and make for easier handling. In addition however, the saw head will tilt at any angle between 0° and 45° so that accurate splayed cutting across the grain can be carried out.

The saw blade teeth on cross-cut saws have what is called a negative hook, which can be seen in the illustrations. The head is also adjustable in height, which permits cutting only

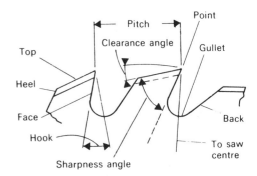

Ripsaw teeth

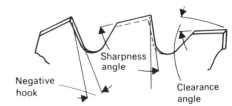

Cross-cut teeth (hardwoods)

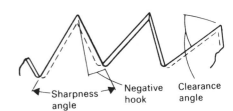

Cross-cut teeth (softwoods)

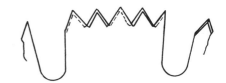

Novelty tooth — for fine cutting with a very clean finish

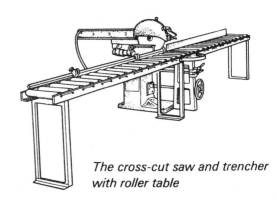

The cross-cut saw and trencher with roller table

part of the way through when required. Although several tables can be used with the cross-cut, the most popular is an all metal table which has ball-bearing rollers to enable long or large timbers to be moved without straining. Accurate cutting-off gauges are fitted to the back side of the table to ensure consistency of length cutting.

When required, the saw blade can be removed and a *trenching head* fitted. This enables the machine to carry out many trenching and jointing operations. When correctly positioned the adjustable front shield protects the operator when in the correct position and the guard can be swung open to give access to the spindle to change blade or fit a trenching head.

In general, however, the ACOP requires cross-cut saws to be used for specific purposes only and not to be adapted.

Narrow band sawing machines

These machines are dealt with under Regulation 11 of PUWER 98, which can be summarised thus:

1   All moving parts, except the cutting section, must be fully enclosed.
2   That part of the saw which is between the top wheel and the thrust wheel must be guarded at the front and one side. The side guard must extend beyond the back of the saw blade.

A *narrow band* saw of the EY-T type made by Robinson is an extremely useful machine to have in the workshop. Apart from being capable of cutting virtually any shaped work speedily and with accuracy it is also useful for cutting regular shapes, square cutting and even small ripping work.

Basically, the saw comprises a frame of strong construction housing an upper and a lower wheel of 760 mm diameter. It is around these two wheels that the continuous narrow band saw blade revolves, passing through the table on which the work is placed. These wheels are called *pulleys* and have a band of rubber around the outer edge like a tyre. The saw blade

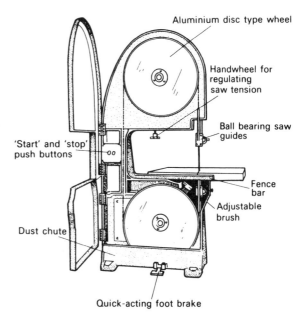

Aluminium disc type wheel

Handwheel for regulating saw tension

Ball bearing saw guides

'Start' and 'stop' push buttons

Fence bar

Adjustable brush

Dust chute

Quick-acting foot brake

The narrow band saw

selected should always be that which is most suitable for the material to be cut. They are available in widths from 6 mm up to 40 mm. In circular work, the smaller the radius of the cut the narrower will be the saw-blade required.

The saw-blade is supported by saw guides situated above and below the table. The guide above the table is adjustable vertically and has a saw guard attached to it. This means that when the guide is set just above the material being cut, there is no danger from an exposed saw edge while working the machine. A *thrust wheel* is fitted behind each guide to prevent distortion of the blade when the work is being pressed against it.

Tensioning of the saw-blade is possible by means of a handwheel situated on the under-side of the upper pulley casing, which raises or lowers the upper pulley to obtain the correct tension. A foot brake located under the lower pulley provides an efficient stopping mechanism as well as being a valuable safety factor.

The table can be tilted up to a maximum of 45° when required: a graduated scale enabling the operator to set the angle accurately. Provision is also made for the use of a fence in any sawing operation. The fence can be used on either side of the saw-blade.

In a machine shop where the output does not warrant the installation of a tenoning machine, the narrow band saw will cut tenons accurately and with reasonable speed. Work can be carried out with the use of jigs or freehand, but the hands should be kept as far away from the revolving saw as practicable in the light of the work being done.

The bandsaw is not confined to cutting timber. Plastic and soft metals can also be cut, but it must be remembered that some metals may require a specially hardened saw-blade.

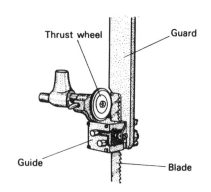

Saw guard showing thrust wheel behind saw blade

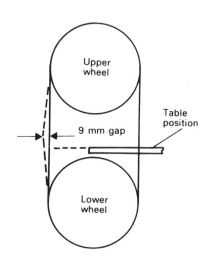

Point at which tension is checked

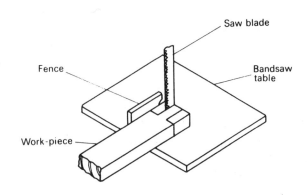

Tenon cutting with fence in position

Planing machines

The ACOP deals at length with safety in the use of these machines but a general summary is outlined, which must be regarded as the minimum standard.

1   The cutters must be effectively guarded *before* work is begun. (**ACOP 11**)

2   Overhand (or surface) planers must be fitted with cylindrical cutter blocks (not square). (**ACOP 4**)

3   A maximum projection of 3 mm is allowed for round form cutter blocks but for other forms the projection can be limited to 1.1 mm maximum. (**ACOP 4**)

4   Overhand planers must be fitted with an easily adjustable bridge guard which completely covers the cutters in such a way as to prevent accidental movement. (**Reg. 4**)

5   When planing the wide face of the material and when planing narrow edges the bridge guard must be positioned as close as possible to the workpiece. (**ACOP 11**)

6   Overhand planers must also be fitted with a guard over the cutters on the back side of the fence. (**Reg. 4**)

7   Push-blocks with handholds must be used when planing short material. (**ACOP 11**)

8   When thicknessing is carried out on a combined machine (i.e. a surfacer and thicknesser) the gap in the table above the cutter block must be effectively guarded. (**ACOP 7**)

9   Thicknessing machines must be fitted with sectional feed rollers or other devices to prevent material being accidentally ejected from the machine, commonly known as kickback. Where such rollers are not fitted, notices must be prominently displayed indicating that single pieces only must be fed into the machine at any time. (**ACOP 24**)

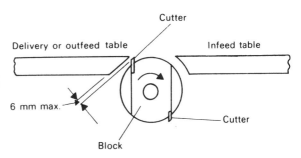

Cylindrical cutter block — showing maximum distance permissible from outfeed table

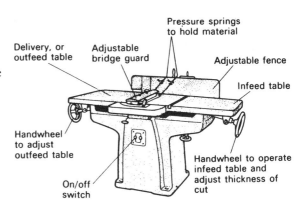

Surface planer

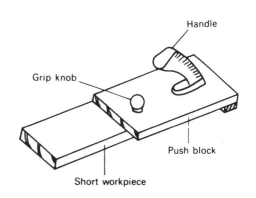

Push block

Planing machines are used in workshops in order to considerably reduce the time taken in preparing timber for use in joinery manufacture. There are many types of machine which will perform this function, but in small or medium sized establishments the more commonly used are:

1   *The surfacer*
2   *The thicknesser*

A thicknesser is a machine that planes material accurately to specified thicknesses. A machine which combines both surfacer and thicknesser is available and will probably be found in smaller firms where neither surfacing nor thicknessing requires to be done at the same time.

The surfacer

Surface planers are designed to plane accurately the face side and the face edge of squared stock. In addition, however, they are capable of chamfering, rebating and preparing boards which have to be jointed along their edges. The actual cutting is done by a cutter-block which, as stipulated in the regulations, must be cylindrical. This cutter-block carries two thin cutters which are rigidly held to prevent chatter. Immediately above the cutter-block is the surface table, which is in two parts:

1   *The outfeed or delivery table*
    This is the part at the rear of the planer farthest from the operator. The outfeed table is only adjusted for sharpening the cutters or for re-setting the machine because the gap between this table and the cutter-block must be as small as possible when in use.

2   *The infeed or feed table*
    Positioned in front of the cutter-block, the feed table is adjusted by means of an easily operated handwheel to suit the amount of cut required for a particular piece of work. It is by lowering this part of the table that rebating up to a maximum of 16 mm is possible. A fence of rigid construction is fitted and can be quickly moved to any position across the width of the table. Pressure springs can be used on the fence for holding down small sections and for other classes of work such as rebating.

It can be tilted at any angle up to 45° and locked in that position to work chamfers etc. An extension to the infeed table enables the fence to be positioned so that the whole width of the cutters can be used where required. On the Robinson type WE-T surfacer this gives a maximum width of 405 mm. In accordance with the regulations the cutters must be fully guarded and this machine is fitted with an easily adjustable telescopic guard. This is called the bridge guard because it spans the cutters like a bridge.

That part of the cutter which is behind the fence (when anything less than full table width is being used) is covered by a guard which is attached to the fence and moves with it when adjusted. This prevents accidents occurring due to someone forgetting to position the guard.

In preparing the face and edge of material on the surface planer it will be necessary to ensure that the widest face is taken 'out-of-wind'. This is the term used when a piece of timber is flat and true after planing, it is 'in wind' when it is twisted and badly shaped.

When truing up in this way it will be found best to push the material across the cutter-block and allow the material's own weight to determine the amount taken off each time it is taken across. When the stock has been passed through several times it will be seen to be lying absolutely flat on the outfeed table, which will indicate that the material face is true and out-of-wind. Once this true face has been achieved and provided the fence has been set at 90° to the table, it will be relatively simple to press the face of the material against the fence and plane a true, square edge. Remember, however, that throughout this operation the guards must be positioned as close to the workpiece as possible.

The thicknesser

The thicknessing machine is the machine to which the material goes when it has received a true surface and edge. It may also be known as a *panel planer*. When the material has been faced and edged the thicknesser will complete the job and bring the material to the required sizes on both its widest face (width) and its narrowest face (edge).

To comply with the regulations a machine which combines thicknessing and surfacing operations must have a *circular* cutter-block. Like the surface machine, the cutter-block on the thicknesser has two cutters.

The material is fed into the thicknesser by what is called the *infeed roller*, which has a fluted face to enable it to take a better grip on the material. This roller is positioned just in front of the cutter-block. Just behind the cutter-block is the *outfeed roller*, which by contrast is smooth all round because the timber is of even thickness by the time it passes through this roller. Both infeed and outfeed rollers are *power driven* and are adjustable when necessary. Situated underneath the feed rollers and in-line with the table are two more rollers which are not power fed. These are called *idle* or *anti-friction rollers*. They can and should be adjusted to suit the moisture content of the timber being planed. Adjustment is possible by a single turnscrew because both rollers are linked to each other. Timber with a high moisture content will require these rollers to be set higher above the table than is the case with drier timber. As will be appreciated from their name, one of the functions of these rollers is to reduce friction and allow the feeding through of material to be controlled and even.

The table is adjusted by a rise and fall mechanism which is operated by a conveniently placed handwheel. A graduated millimetre scale gives an accurate guide to thicknesses required. Remember that inaccuracy may occur if the reading is not taken in line with the pointer on

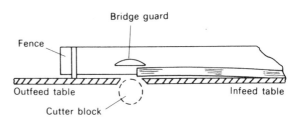

Truing a bad shaped board to form a face side

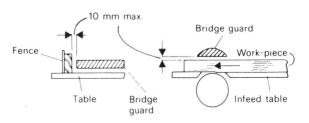

Position of bridge guard when edging and facing materials

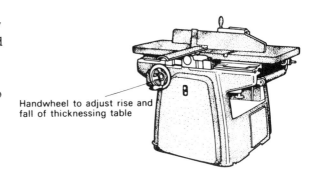

The thicknesser — this model combines surfacer and thicknessing

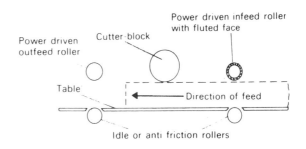

Relative positions of cutters and rollers with material being fed through

the scale. The sequence of operations should be to carry out the widths first and if there are various thicknesses the thickest stock should be planed first, working down to the thinnest pieces.

Do not take too much material off in one pass and *never* more than 3 mm at one go. On the other hand, ridge marks caused by the infeed will occur if less than 1 mm is removed. Dust extraction equipment is required by the regulations. This is extremely important with the thicknesser because the chippings which are produced whilst planing must be cleared immediately if bruising to the face of the material is to be avoided. Combined thicknessing/surfacing machines are exempt from this regulation. Remember, however, the regulations do insist that on combined thicknesser/surfacers the cutters on the upper side must be adequately guarded.

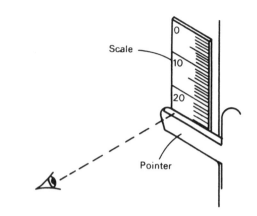

Take reading of scale in line with pointer

Mortising machines

Although mortising machines are not specifically referred to, the ACOP requires the following measures to be observed at all times:

1   All cutters must be enclosed by substantial guards to the greatest possible extent.
2   No guards to be adjusted while the machine is in motion.

Machine mortising is carried out in two ways:

1   *Chisel mortising* is when the mortises are cut by the combination of a hollow chisel inside which an auger bit rotates, cutting away the bulk of the waste.
2   *Chain mortising* is a method of cutting the mortise by means of a rotating chain.

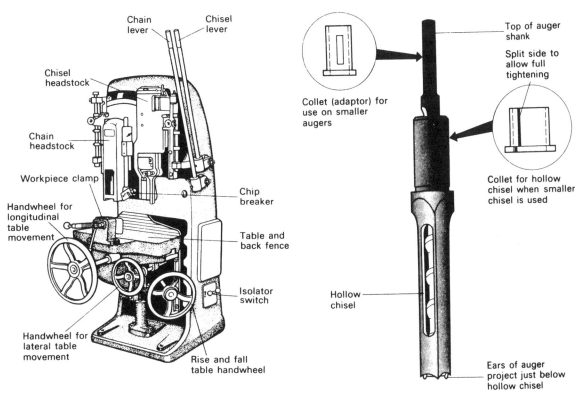

Combined chisel and chain mortiser

Hollow chisel and auger

Some machines are designed for either chain or chisel. Others have single spindles on which can be fitted a chisel or chain headstock. The Robinson SL-E mortiser has both headstocks situated side by side. They can be used singly or in conjunction with each other. Each headstock is operated separately by an adjustable hand lever located on the right-hand-side of the machine and both have dual depth controls to enable the cutting of haunch and full mortise at the same time.

Chisel mortising

The hollow chisel section will cut mortises from 6 mm up to 25 mm in softwoods and up to 19 mm in hardwood. First, the appropriate size of chisel is selected and the hollow chisel portion is secured in the chisel headstock. After this the auger is passed through the hollow core

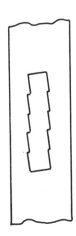

The cutting effect of a chisel which is not fitted square to the fence

into the chuck, ensuring that the ears of the auger are projecting just below the chisel. This is necessary to give clearance between the auger and chisel during use. Adaptor bushes (collets) are used when fitting the smaller augers into the chuck, which is made large enough to accommodate the 25 mm auger. During use air is blown through a nozzle to keep the work free from chippings.

Care must be taken to square the chisel with the back fence, otherwise irregular mortising will occur. Hollow chisel mortising is the more accurate method and is, therefore, used more on 'one off' jobs or work where a high quality of finish is demanded.

Chain mortising

This method is considered much more appropriate when there is a large quantity of mortising to be done. Chain sizes range from 6 mm to 32 mm wide, for both soft and hardwoods. While the chain is in use an exhaust fan draws off the chippings to keep the work clear. A guard is fitted which has a safety glass panel to give the operator a full view of his work.

A chipbreaker is a feature of the chain mortiser. Often made from hardwood it is adjustable and is situated on the exit side (right-hand-side) of the mortise to prevent breaking-out taking place. With the continuous rotation of the chain, breaking-out is a constant possibility. Chain mortising is much more rapid in execution because more timber is being cut away at any given time. It must be remembered that with the smallest chain the mortise length will be a minimum of 20 mm and with the largest chain, i.e. 32 mm, a minimum length of 37 mm. This contrasts sharply with the hollow chisel which is square.

The table on this mortising machine is precision ground for accuracy and is

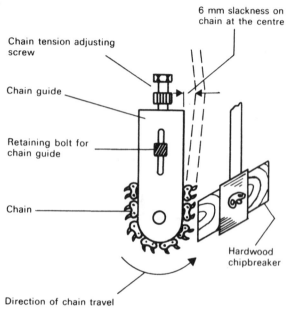

Mortise chain and chipbreaker

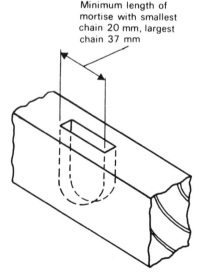

Restriction on mortise length when using chains

approximately the same width as the machine itself. At the front and in the centre of the table is an easily operated, quick action clamp which holds the stock securely in position during mortising. The table can be adjusted upwards or downwards by the operation of a conveniently placed hand-wheel. In addition, there is adjustment forwards or backwards horizontally which enables the work to be accurately set in line with the chosen headstock (i.e. chain or chisel) before commencing cutting. Having fitted the correct chisel or chain for the job and also correctly positioned the headstock in relation to the work the material should then be placed and clamped on the table.

Longitudinal movement up to 700 mm can be achieved by means of the large hand-wheel in the front, but where mortises are further apart, such as on door stiles, it will be found necessary to re-position the timber to carry on mortising. Whether using chain or chisel it is best to commence cutting the mortise in a central position and gradually work towards the setting-out lines on each side.

The effect of alternating the material face against the fence with off-centre mortises

Mortising by machine, as with hand mortising, should be done from each side, cutting through approximately half way and then turning the stock to complete the mortise from the other side. For this reason it is important to keep the face side of the material against the back fence of the table at all times, particularly if the mortise is not placed centrally in the material thickness. Failure to observe this simple rule will result in mortises running inaccurately through the timber and causing framework to twist in assembly.

Woodworking lathes

Once again, no specific mention is made to the use of lathes in the regulations, but the following general principles apply:

1 Turning tools must be kept sharp, as blunt tools contribute to accidents.

2 No packings should be used as they will gradually become compressed and loose.

3 Carry out a check of securing bolts or heads before using the machine.

Lathes for woodturning are found in a variety of designs and sizes. In this section we shall concern ourselves with the *hand woodturning lathe* and not the automatic lathes which are to be found in mass produced furniture factories where much repetition turning is required. As woodworking machines go, the lathe is a comparatively safe machine to use. It is the timber being turned which spins, while the tools are held against the hand and tool rest. The tools are used to remove unwanted stock and form the required shape.

The turning speed of the lathe motor ranges from as little as 200 rpm to as high as 3000 rpm In many models this is achieved by pulley wheels of various diameter to give the appropriate rpm. Turning pieces of large diameter or section would normally be carried out at the lower speeds, while smaller diameters are turned with the lathe revolving at the higher speeds.

The Myford ML8 is a modern lathe which features all the characteristics of the older traditional models. At the left-hand-side is found the motor and *headstock*, on each side of which is a *faceplate*. The *tailstock* is found at the opposite end, with the *hand and tool rest* between the two. Both tailstock and tool rest are adjustable to enable correct and safe positioning. While the machine is in use tools can be placed on the shelf just below the position of the work. This lathe will take work which is up to 1.065 m in length.

For safety reasons lathe motors are normally housed and in this model the motor is housed just below the headstock in a cupboard which can also be used for storage purposes.

Woodturning is carried out in two ways:

1 Turning between centres: this is when the timber to be turned is secured by the headstock and tailstock while being shaped.
2 Face plate turning: such as fruit bowl turning, where the block from which the work is to be turned is secured to the face plate only.

Various centres are used for the different shapes

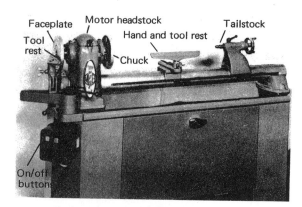

The Myford ML8 woodturning lathe

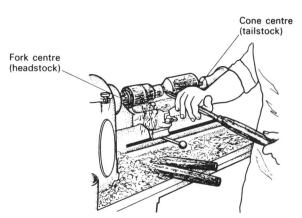

Turning between centres

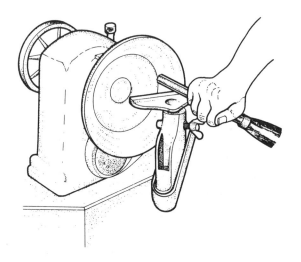

A bowl being turned using a faceplate

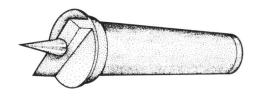

Fork centre (headstock)

Cone centre (tailstock)

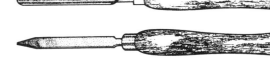

Faceplate (left) and screw chuck

Gouge (top) and parting tool (bottom)

required, but the most universal are the *fork centre* for headstock and the *cone centre* (revolving centre) for tail stocks.

Woodturning tools

The tools used in woodturning consist of chisels, gouges and scraping tools. These tools have similar uses to those which are used in traditional joinery or woodwork operations. However, the chisels are ground on both sides and the skew chisel also has a splayed cutting edge. The other notable difference being the stronger blades and longer handles which are a feature of turning tools.

Gouges are used to cut hollows and *scraping tools*, which can be ground to suit any required shape, are the main finishing tool, followed by fine glasspaper. The *parting tool* separates the finished work from the unwanted amount at the ends into which the stock centres are inserted.

During any turning work a check should be carried out to ensure that the material is still firmly held. Not all timber is suitable for turning and, therefore, care should be exercised in timber selection. When setting the work up

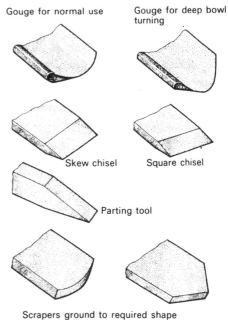

Gouge for normal use   Gouge for deep bowl turning

Skew chisel   Square chisel

Parting tool

Scrapers ground to required shape

Various turning tools

the tool rest should be positioned as close to the material as possible and the wood should be spun by hand to make certain that it will clear the tool rest. On turning work as much removal of waste as possible should be carried out beforehand. For instance, before a piece of square material is turned the corners should be removed so that the piece is octagonal when secured to the head and tailstock.

Preparation of square material before turning

PUWER 98 Regulation 4

This regulation is concerned with the suitability of work equipment. To comply with the ACOP every employer must ensure the following requirements are met:

Extract equipment
A suitable and effective extraction and dust collection system must be installed and fitted to the following machines:

1   Thicknessing machines
2   Vertical spindles
3   Multi-cutter moulding machines
4   Tenoning machines
5   Automatic lathes (not hand lathes)
6   High speed routers

Lighting
Adequate and suitable lighting must be provided in all machine shops and direct rays of natural light must be shielded to prevent them dazzling a machine operator.

Noise
Ear protectors must be provided in shops where persons are exposed to noise levels of 90 dB (decibels) for eight hours per day or more.

Machine shop layout

Bearing in mind the requirements of the ACOP in the safe use of woodworking machinery, the following recommendations are made:

1   Lay out machines in operational sequence
2   Provide sufficient back space for each operator
3   Ensure timber being machined does not interfere with other machines
4   Keep passage and walk-ways clear
5   All material not in use should be properly stacked and clear of machines

Careful thought must go into the positioning of machinery in the workshop. A sensibly laid out shop will not only achieve a greater degree of safety, which is very important, but will also enable production costs to be kept to the minimum. This will be true whether the machine shop is so small that the joiners' work benches are also situated in the same area, or where the joiners shop and the machine shop are entirely separate units.

Of course, one of the most important factors in the layout of a machine shop is the number and type of machines which have to be accommodated. In very small firms it may well be that only one circular saw bench and a combined surfacer/thicknesser will be sufficient for that Company's needs. Naturally, the larger the output the more machines will be necessary to carry out the various operations.

Also of great importance is the position of the timber stack in relation to the machine shop, especially if the timber stack already exists. Ideally, the timber stack needs to be fairly close to the workshop entrance and if it does not already exist this is where it should be sited. In looking at the functions of the various machines it will be appreciated that the crosscut saw will also need to be positioned close to the entrance and, therefore, near to the stack to minimise the distance that materials are moved before being brought to length.

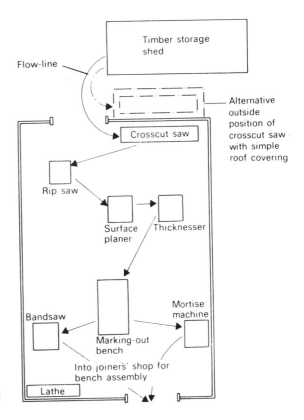

Lay-out of a small machine shop for economic handling of material

The crosscut saw is often positioned inside the building, but in cases where there is a shortage of space the machine can be set-up outside the machine shop under a simple lean-to roof to give some protection. Following the crosscutting of material it must be borne in mind that machine operations follow exactly the same sequence as hand operations. This really means that the timber will be faced and edged before being planed to the width and thickness required on the thicknesser. From this point the material is marked-out on the *marking-out bench*.

As with the hand method, mortising will be completed before tenoning is carried out. Where no tenoning machine is available these

tenons can be cut quite adequately on the band saw. It is with this *sequence of operations* in mind that machine shop layouts will be planned. The name given to the path through which the materials pass on their way to completion is the *flow line*. Flow lines play an important part in the planning of all workshops and should assist in speedier production and lower costs.

There is no easy formula for the successful planning of machine shop layout, but the answer will lie in giving careful thought to all the considerations mentioned here, as well as others which may present themselves in individual cases.

Questions

1   What is the ACOP? Describe its function.

2   What is the intention of Regulation 9 in PUWER 98?

3   What is the function of the riving knife on a circular saw?

4   Where is the crown guard to be found and by what name is it also known?

5   List and describe the aids used in conjunction with circular saws.

6   Where is the thrust wheel situated on a narrow band saw and what is its function?

7   Explain the reason for limited projection of cutters. What is the maximum projection allowed on a simple form round cutter block?

8   Describe the function of anti-friction rollers on a thicknessing machine.

9   Describe the types of work most suitably carried out by (a) chain mortising and (b) chisel mortising.

10   Using a sketch, describe the positioning of machinery in a workshop to give maximum efficiency.

11   What is meant by the term 'flow line'?

6 Doors, door frames and linings

Doors and their frames take up a great deal of the working time of carpenters and joiners, whether making them in the joiners' shop or fixing them on site carpentry work. The woodworker will need to know about the quality and types of door in use today, including methods of construction. The quality of a door will determine its serviceable life and that is why two categories are made:

External quality: This category includes all doors which are to be used as front or rear entrance doors. Such doors are likely to be exposed to weather conditions for most of the time and need to be constructed in such a way as to enable them to stand up to it.

Internal quality: Most doors will fall into this category because very many more doors are used inside than are used externally.

Probably the biggest single factor affecting external and internal doors is the type of adhesive used. It will be appreciated that in the construction of external doors an adhesive must be used which will have very high resistance to moisture and dampness. If this is not so, then all joints will begin to deteriorate very shortly after the door has been put into use.

There is a very wide variation in both size and style of door. Although a range of standard sizes is available, it will be understood that in special circumstances doors can be manufactured to suit any size or requirement. It is also usual for the external quality door to be thicker than the internal door. In most cases the exterior grade is 44 mm thick (referred to as 2″ nominal) because they are made from standard 50 mm (2″) thick material, while interior grade is 35 mm (known as 1 ½″) material is used. It is necessary here to use metric sizes with imperial equivalents because many stockists have yet to change to metric measurement.

The standard sizes which have been in use for so long are:

2134 × 914 mm (7′ 0″ × 3′ 0″)
2032 × 813 mm (6′ 8″ × 2′ 8″)
1981 × 762 mm (6′ 6″ × 2′ 6″)
1930 × 711 mm (6′ 4″ × 2′ 4″)

Although it is possible to make doors to any design, those which have been in most common use are:

1 Ledged and braced doors and framed ledged and braced doors
2 Panelled doors
3 Flush doors
4 Louvred doors

There is a marked difference between the methods of construction of mass produced doors and those made in workshops where mass production machinery is not available. For instance, many factory made doors are dowelled together instead of being mortised and tenoned.

Ledged and braced, and framed ledged and braced doors

These two types of door have traditionally been used for external purposes such as on outbuildings, sheds etc. The ordinary *ledged and braced* door is also used in close boarded fencing as a gate.

Ledged and braced doors
Construction is of the simplest kind, consisting of tongued and grooved match-boards, usually vee jointed (TGV), and held together by ledges. The ledges are fixed by nails across the top,

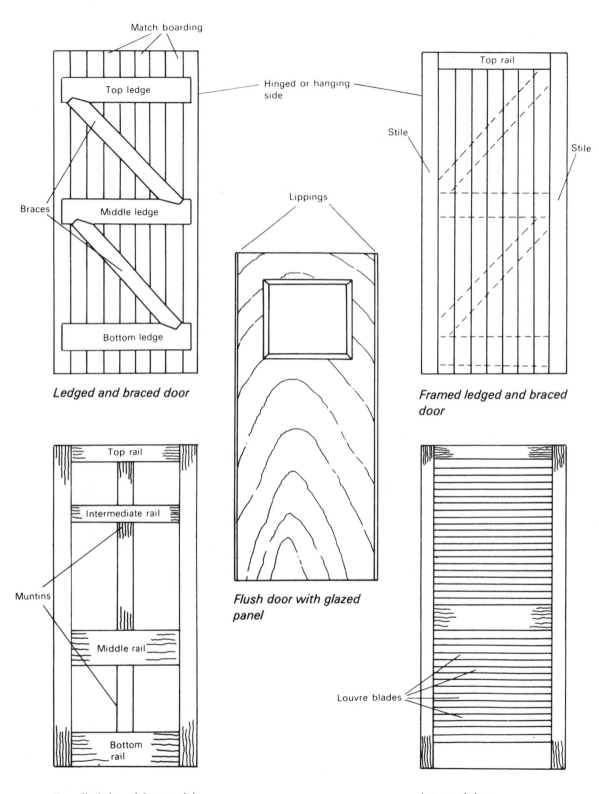

Match boarding

Top ledge

Hinged or hanging side

Braces

Middle ledge

Bottom ledge

Ledged and braced door

Lippings

Flush door with glazed panel

Top rail

Stile

Stile

Framed ledged and braced door

Top rail

Intermediate rail

Muntins

Middle rail

Bottom rail

Panelled door (six panels)

Louvre blades

Louvred door

middle and lower parts and strengthened by diagonal braces of similar thickness. Braces must be fixed with the top being positioned on the closing side farthest from the top hinge and running diagonally down to the bottom where the brace is close to the bottom hinge. This will prevent any drop taking place due to the weight being placed on the closing side. Because of this the doors have to be made to suit right- and left-hand hanging requirements. To determine the hand of any door it should be viewed from the side where the hinges are seen (the inside). If the hinges are on the left the door is left-handed and if on the right it is right-handed.

The match-boarding is normally 100 × 15 mm, but the ledges and braces must be of stouter material and not less than 32 mm thick. No adhesive is used on this type of door, but good quality wood primer should be applied to the tongues and into the grooves before assembly. The same treatment must be given to the back face of the match-boards and underneath the ledges and braces to ensure that these parts are given protection against the weather. It will be noted that a small splay is applied to the top edge of the ledges to ensure that water or dampness runs off and prevents decay from setting in. There are no conventional joints used in the ledged and braced door, the matchboards are simply nailed through the face into the ledges and braces. The joints between the diagonal brace and the ledges will be stronger if they are constructed as shown in the illustration.

Framed ledged and braced doors

Framed ledged and braced doors are also constructed of TGV boarding with ledges and braces, but with the addition of two stiles and a top rail to give greater strength. Haunched mortise and tenons are used to secure the joint between the top rail and the stiles. The middle and bottom ledges have bare faced tenons which

Determining the hand of door hanging

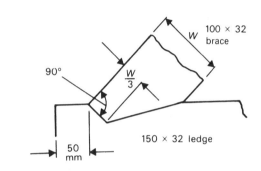

Detail of joint between ledge and brace

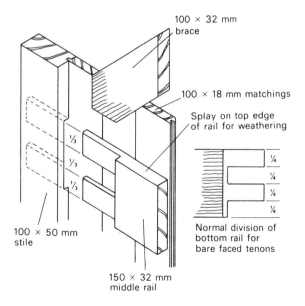

Detail of joint at junction of brace and middle rail (showing bare-faced tenon)

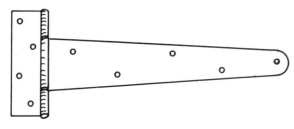

Scotch tee hinge

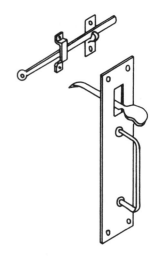

are mortised into the outer stiles and the match-boards are tongued into the stiles as well as the underside of the top rail. In this door the mortise and tenoned joints should be glued with an exterior grade adhesive, but as with the ledged and braced door all other parts should be thoroughly primed before assembly.

Suffolk latch

Ledged and braced doors are usually hung with *scotch tee hinges* and fitted with a *suffolk latch*, whereas framed ledged and braced doors are fitted with *butt hinges* and *rim lock* with *rim furniture*.

Panelled doors

These doors are made for internal and external use with as many or as few panels as required. They are often identified by the number of panels, e.g. two, three or four panelled door.

The names of the various parts of a panelled door are the same as for any other door. *Stiles* are the two outer vertical pieces which gives the door its edge. One is on the side on which the hinges are screwed and this is called the *hanging stile*. The other stile is the side on which the lock and other catches are fitted and is called the *closing stile*.

Rails are the members which cross the door horizontally. The *top* and *bottom rail* are present in all doors and in most cases the bottom rail is wider than the top. Other rails

can be included, depending on the number of panels required. If a rail occurs in the centre of the door this is called the *middle rail* and is often wider than the stiles. It should be noted that this rail is not truly in the middle, but will be slightly nearer to the bottom of the door. This rail is also frequently called the *lock rail* because the lock is often fitted to the door at the same height from the floor.

Muntins are used when the width of the door is to be sub-divided. Like the stiles they are vertical members, but the muntin is never carried through in full length, it is always cut between rails.

Materials used for the panels vary according to the quality of the door. In good quality

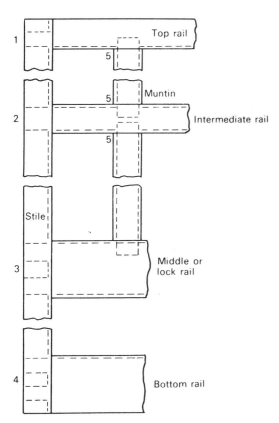

Joints

1 Top rail — haunched M and T into stile
2 Intermediate rail — plain M and T into stile
3 Middle rail — pair of through tenons
4 Bottom rail — pair of haunched tenons
5 Muntins — stub tenoned into rails

*Arrangement of joints in a panelled door —
dotted line represents panel groove*

Plain groove   Rebate and bead   Planted beads

(*Note*: beads must
be nailed to door
and not panel)

(*Note*:1.5 mm gap in panel to allow for expansion)

Methods of securing panels

Bolection moulding
screwed from behind

Planted bead fixed after
bolection moulding

Raised and fielded panel with bolection moulding

hardwood doors it is possible that the panels will be made from solid timber. However, plywood is much more frequently used nowadays and hardboard is also used in cheaper quality doors.

Methods of securing panels

Grooves
This method is used extensively. The panel is grooved into stiles, rails etc. on all its sides. The depth of the groove should be made the same as the panel thickness, up to a maximum of 16 mm.

Rebate and bead
This is another common securing method. The beads can be plain, chamfered or moulded as required and are accurately cut in. Some beads can and should be mitred, whereas others may have to be scribed because of their design.

Beaded panels
These are not so frequently used. In this method, planted beads are fixed on both sides of the panel. It is not as satisfactory as the other two methods, but in some situations it provides the only method. Many panels are left plain, but in some good class work they can be raised and fielded to give a better appearance. A bolection mould is often used with raised panels. This mould is so designed to fit neatly over the rebate formed by the panel surface with the stile of the door.

As mentioned earlier, the joints used in some mass produced doors are dowelled and this is true of some panelled doors. However, those which are not mass produced will be framed up with mortise and tenon joints. It will be noted that the wider middle and bottom rails are cut

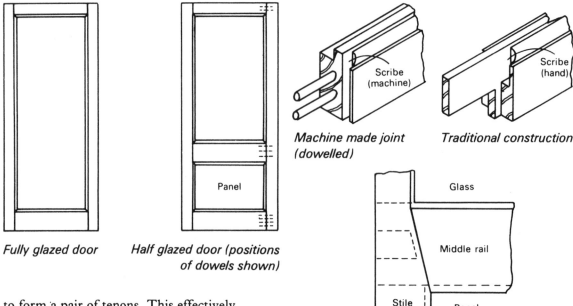

Fully glazed door | *Half glazed door (positions of dowels shown)*

Panel

Machine made joint (dowelled) | *Traditional construction*

Scribe (machine) | Scribe (hand)

Diminished stile

Glass | Middle rail | Stile | Panel

to form a pair of tenons. This effectively reduces the width of the rail and therefore lessens the chances of the rail twisting or bowing.

Muntins are stub tenoned into the rails and never tenoned right through.

Glazed doors

Fully glazed and half glazed doors are also used extensively and are constructed in a similar way to panelled doors. In most of these doors the glass is placed into rebates and loose beads are cut and fixed to secure it. In addition, wash-leather tape should be attached to all edges of the glass to act as a shock absorber when the door is slammed. Putty, although used to secure glass in windows, is not a suitable material for glass panelled doors.

In some half glazed doors the stile below the middle rail is wider than that above. This increases the size of the glazed area above the middle rail to allow the passage of more light. As a result the door has what are called diminished stiles, which are illustrated. Doors of this type of construction should be assembled, glued and wedged-up on a bench

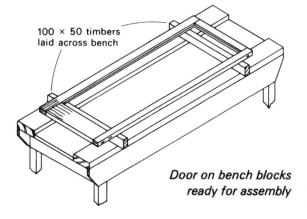

100 × 50 timbers laid across bench

Door on bench blocks ready for assembly

which is free from all other materials. Two bearers of 100 × 50 softwood should be placed across the bench near to the top and bottom of the door. These bearers must be checked to be out-of-wind before commencing to ensure that the door when assembled is true. Wedges must be glued and driven in methodically, driving the outer wedges first. Cramp positions must not interfere with the wedging process.

Flush doors

These are so called because both sides of the door are covered with a continuous flat panel. The panels are usually plywood, but in cheaper mass produced doors hardboard is also used extensively. Both long edges have strips of wood glued on to mask the ply or hardboard edges. These strips are sometimes called *edgings*, but are more often known as *lippings*. Various methods of lipping flush doors are used, the commonest being:

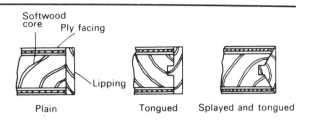

Common methods of lipping edges of flush doors (long edges only)

1   *Plain lipping* is very common and consists simply of a thin strip 6–10 mm thick being applied to both long edges of the door.
2   *Tongued and grooved lippings* are an improvement on the plain lippings because they form part of the softwood inner framework.
3   *Splayed and tongued lippings* are used in situations where it is desirable not to have the lippings showing on the faces of the door. This would apply, for instance, on a flush door with hardwood veneered faces which would be spoiled if the thickness of the lippings were seen from the face.

Core construction

With the exception of fire resistant doors which will be dealt with later, the vast majority of flush doors are hollow core. Once again, there are two quite distinct methods of core construction. Most of the mass produced doors have a softwood frame around the top, bottom and sides, while the central area consists of a form of honeycomb patterned cardboard which resembles egg crates and is often referred to by that name. Better quality doors have a more substantial softwood core framework which has horizontal as well as vertical pieces to hold the outer panels rigid. Whatever type of core the door may have, all flush doors must contain a *lock block*. This is a piece of solid timber placed and glued into the position where the lock will

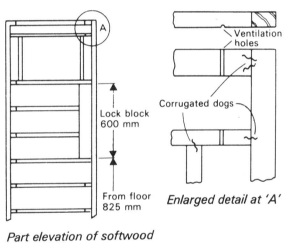

Part elevation of softwood core framing showing lock block and glazed panel

Enlarged detail at 'A'

Corrugated dog

be fitted. The block must be clearly marked on the finished door to assist the operative when fitting and hanging the door and to ensure that the door is hung on the side farthest from the lock block. The joints used on these cores are of the simplest kind. Butt joints are quite normal, with either dowels or corrugated dogs to hold the framework together while the outer skins are glued and pressed on.

When constructing a flush door with an all timber core it is necessary to include ventilation holes which will ensure a continuous air-flow over the whole of the core. Absence of these vents may cause the air trapped inside the core to stagnate and result in distortion or movement in the face panels.

Observation panels can be incorporated into flush doors. This will mean that softwood framing will need to be constructed outlining the glass size. The glass can be secured by means of plain planted beads or rebated beads which can be mitred neatly and pinned into position.

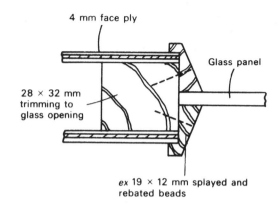

4 mm face ply

Glass panel

28 × 32 mm trimming to glass opening

ex 19 × 12 mm splayed and rebated beads

Note: Rebated beads mask the edge of face ply

Method of securing glass panel in flush door

Louvred doors

In recent years louvred doors and panels have been used increasingly for purely decorative reasons, such as fronts for fitted wardrobes or cupboards and kitchen units. Originally, louvres were used in door manufacture to provide ventilation where no other form of air-flow existed. The motor room of a lift on the roof of a multi-storey building would often have louvred doors to prevent build-up of heat inside because no other windows or openings existed.

Construction
Louvre blades are inclined at an angle of 45°, but can be greater or less than this to suit certain requirements. The width of the louvre blade will depend on the inclination, but it will always be greater than the overall thickness of the door. Thicknesses will vary, although in the decorative door they seldom exceed 10 mm. Because of an increase in demand a very large quantity of louvred doors are totally machine-made and the joints are dowelled together. However, a better quality will be achieved if all joints are mortised and tenoned as in the conventional method of door construction. The

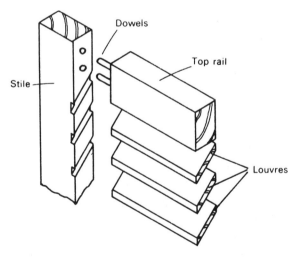

Dowels

Top rail

Stile

Louvres

Construction detail of louvre door

stiles and rails should be set out in the normal way, and at the same time, the housings for the louvres should be marked on to the stiles before any work is started.

Door frames and linings

Depending upon the position in which it is used the timber surround to the door will fall into one of the following two categories:

1 Door frames
2 Door linings

Door frames

Door frames are of substantial construction having mortises and tenons on the *heads, jambs* and *sills* (where used) and in many cases, having the stops rebated from the solid timber. Chamfers or mouldings are sometimes applied to these frames. To give additional strength to the mortise and tenon joint on this type of frame wooden dowels of about 12 mm diameter are inserted through the centre of the tenon face from one side only. These dowels need not go right through the whole thickness of the frame provided the dowel goes through the whole tenon thickness and part way into the frame beyond.

A method called 'draw boring' is used to ensure that the shoulder fits very tightly to the rail. Firstly, a hole is bored into the frame. A hole is then bored into the tenon, slightly off-set from the hole in the rail. When the dowel is driven in it will pull against the tenon to produce a joint that is fully cramped.

Door linings

Door linings, which are of much lighter construction than door frames, are used exclusively for internal doors. Linings usually have planted (nailed on) stops and, in the

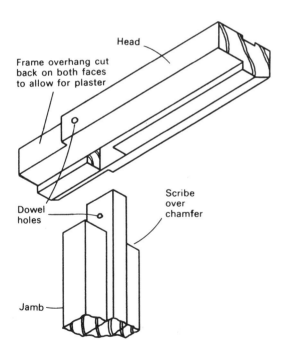

Detail of joint between head and jamb of frame

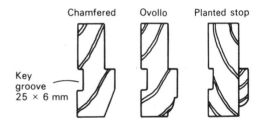

Various frame sections

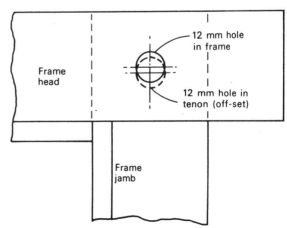

Method of boring holes when draw boring

majority of cases, have a tongued housing joint between the jamb and the head. Door linings seldom have sills.

The heads can be allowed to run past the jambs on each side for about 100 mm or so. This gives additional strength when it has been built in, but it will be noted that the face is cut back on the oversail, equal to the plaster thickness, so that it is not seen in the finished work. Frames are always treated in this way, but it is not always possible in the case of linings.

Types of frames and linings

Many types of frame or lining are made, some with a glass panel above the door. This is often done internally to give extra light to a part of the building where no windows exist, such as at the top of the stairs on the landing in a house. A frame or lining performing this purpose is called a *borrowed light* frame or lining and the rail separating the door from the glass above is called the *transom*. Those which extend from the floor to ceiling are called *storey frames or linings*, these also include the transom rail. Whenever possible door frames are built in as the building progresses. Linings are sometimes fixed in this way, but they are more commonly fixed after the brick or block work has been completed.

Before leaving the joiners' shop all frames and linings should have cross bracing and diagonal bracing to keep the frame or lining rigid at all times. These braces must not be removed until after the frame or lining has been securely fixed. Frames that contain sills need only a diagonal brace. Some, but not all door frames have a sill which will normally be hardwood because of the extra wear placed upon it by being walked on. Where no sill exists it is quite usual to find a mild-steel dowel of about 12 mm diameter inserted into the bottom of each jamb, projecting about 32 mm. The dowel is bedded into the concrete to give a firm hold to the foot

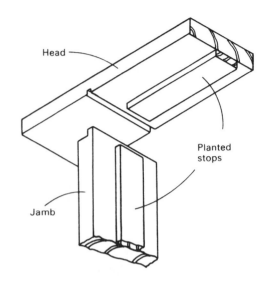

Tongue housing joint between lining head and jamb

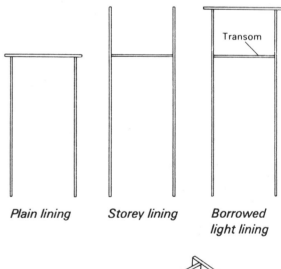

Plain lining   *Storey lining*   *Borrowed light lining*

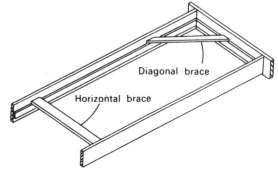

Braces fixed to lining before delivery to site

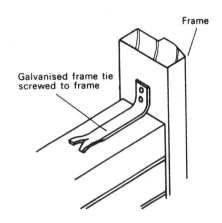

The use of frame ties for building-in frames

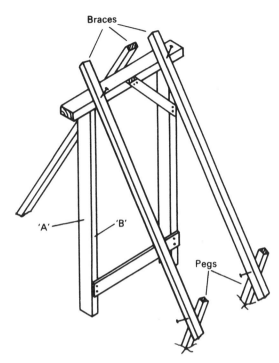

(*Note*: Faces 'A' and 'B' must be plumb)

Frame held by braces nailed to a peg driven into the ground

of the frame. Reinforcing rod is ideal for this purpose and there are usually some odds and ends lying around for use.

Positioning frames and linings

Frames should be stood and braced in position before brickwork is started. A variety of methods can be used to do this, but whichever method is used the main question is: 'Will it move during construction?' If it is likely to move, then it must be more securely braced. Time should be spent making certain that the frame is plumb (vertical) both on its face and on its side. When this is done satisfactorily, battens should be fixed diagonally to hold it accurately in position while bricklaying is carried out. As the brick courses proceed, galvanised frame ties are screwed to the back side of the jambs and

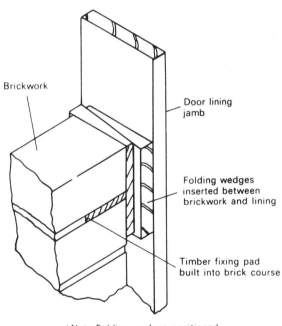

(*Note*: Folding wedges positioned immediately behind fixing pad)

Method of fixing linings

built in with the brickwork. The ragged ends of the frame ties provide a very strong fixing for the frame.

As the brick or block work proceeds internally fixing pads of timber are inserted approximately

600 mm apart to which the linings are screwed or nailed. The opening size should be slightly oversize so that folding wedges or packings can be inserted behind the lining jambs. This gives an opportunity to keep the jambs of the lining absolutely straight and parallel. The thinner jambs of the lining could easily distort and cause great problems when the door is hung.

Architraves

When door frames or linings have been fixed and plastering has been completed there is a joint between the lining or frame and the plaster surface at the sides and above the door. This joint is covered by a timber surround called an *architrave*. These can be plain or moulded, large or small. There are a number of standard patterns in use which include splayed and rounded, ogee and plain rounded.

Architraves are cut with a mitre (a 45° splayed cut) at the intersecting angles and fixed to the door frame or lining with oval nails which should be punched in below the surface.

Architraves are fixed to the frame or lining, leaving a small parallel distance between the edge of the frame or lining and the edge of the architrave, this is known as the *margin*. There is no set distance of margin.

Plinths are sometimes fitted at the foot of the architrave at its junction with the skirting. It can make a more elegant job of finishing the skirting boards. They can be plain or splayed, or any desired shape, and are usually terminated just above the height of the skirting. Plinths, too, are fixed by nailing.

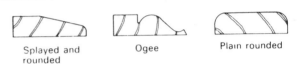

Splayed and rounded   Ogee   Plain rounded

Types of architrave in common use

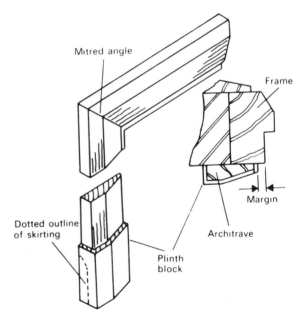

Details of architrave fixing

Door hanging

Doors are delivered to site in bulk before they are required. During the period from delivery to the time they are to be hung they must be stored very carefully. This means that correct stacking is necessary and they must be fully protected from weather. The only certain way to do this is to lay the doors flat with battens of equal thickness between them, just as timber is stacked for seasoning. In this way the doors will not only be adequately ventilated, but their own weight will ensure that they remain flat and out-of-wind until they are required for use. Doors which are stacked in the open must have a heavy duty tarpaulin or similar covering to prevent soaking.

The importance of accurately fixed frames or linings will be realised when the door is being prepared for the opening. To commence the door hanging operation two carpenter's stools of similar height should be spaced in such a way that the door can be laid flat across them with the face or best side uppermost. If there is a side of the door which is of poorer quality the door should be hung with the best face in the most advantageous position, i.e. on the room side where it will be seen more often.

With the door laid flat on the stools the *horns*, which are the over-running stiles left on for protection, are first removed from the top of the door. The top is planed with a jack plane to make it straight. A pinch rod, which need only consist of an off-cut about 18 × 12 mm, is then placed vertically in the door opening from the floor and marked at the head, making an allowance for the floor covering thickness. This is then checked across the full width of the opening to ensure that it is parallel. The height of the door is now ascertained and should be marked onto the door, measuring from the head which has been previously trued up. The surplus material is cut off and planed to a finish.

The next stage is to shoot one of the door edges accurate and straight. The pinch rod is then used to mark the width of the door, making checks at various positions in the height to ensure that it is parallel. Allowance must be made for the joint around the door. The joint should be 2.5 mm, which means that the door width must be finished 5 mm narrower than the actual opening size.

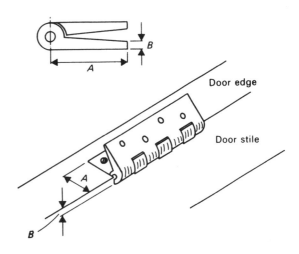

Cutting in the hinges

Cutting-in the hinges

With the door now brought to the correct size the *butt hinges* should be let in, on the hanging stile first. The top of the top hinge should be placed 150 mm down from the top of the door and the bottom of the bottom hinge should be 225 mm up from the bottom. The hinge flap must be let-in equally on the door and frame, or lining. When the hinges have been screwed to the door it is best to offer the door into position, using a wedge underneath to hold it steady, while positions can be marked for the hinges. The door is then removed while the hinges are chopped in to the frame. The door can now be hung into the frame or lining.

Fitting a mortise lock

To fit a mortise lock to the door the following sequence of operations should be followed:

1 Wedge or secure the door in a half-open poisiton.
2 Accurately mark the position of the mortise on the edge of the door and also the position of the spindle and keyhole on the faces.
3 Bore holes for the lock spindle and keyhole.

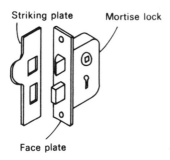

Upright mortise lock

4 Using a twist bit slightly larger than the thickness of the lock body, bore a series of holes in the centre of the door edge.

5 Clear waste material with a mortise chisel and place the lock in the mortise.

6 Mark face plate of lock and cut-in accurately.

7 Screw lock in position and close door sufficiently to give position of latch from which to position striking plate.

8 Mark-out and accurately cut-in the striking plate. The door furniture, i.e. handles etc. is usually fitted after the paintwork is completed.

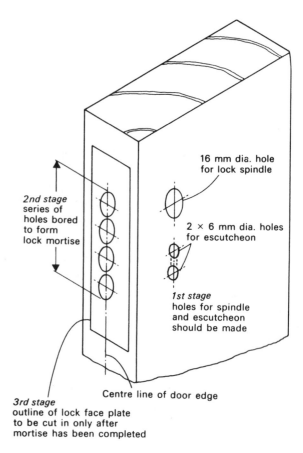

The sequence of fitting the mortise lock

Door sets

All doors, frames and linings so far referred to have been of traditional construction. *Door sets*, which are now used quite extensively, represent a new concept. They are for internal use only and have been developed to cut down the amount of on-site work in fixing doors and frames. They also reduce costs. Door sets are entirely pre-assembled, with doors hung and locks fitted etc. and are fixed on site after the plaster work has been completed. All sizes in the range are metrically co-ordinated for replacement and inter-change purposes. Plywood profiles equal to the overall size of the door set frame are placed in position so that the plasterer is given an edge to finish to. The profile is removed when the door set is fixed.

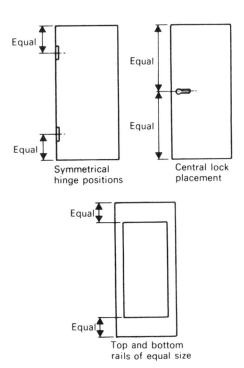

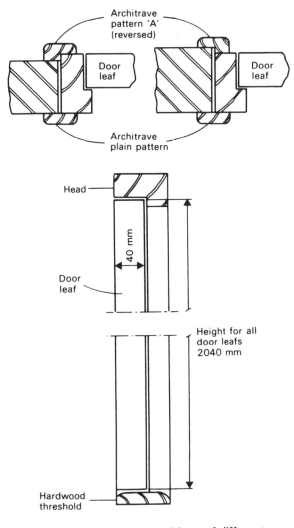

Common features of internal door sets

Door set frames are made of softwood to suit various partition or wall thicknesses. The thickness of the frame is usually slightly less than the traditional frame thickness, but thicker than the traditional lining thickness. Frames can be door or ceiling height. Ceiling height frames have a glass panel above, as in the borrowed light frame or lining. Thresholds (or sills) are optional, but when required are 15 mm thick and in hardwood. Machine combed joints are used between the head and jambs of the frame and there is no over-run. The transom, when used, is tenoned into the jambs.

Doors are of flush faced hollow core construction, 40 mm thick, and can have a paint or hardwood veneer finish. They are hung into the frame with specially developed *'snap-in'* hinges. These are spaced equally from top and bottom so that doors become inter-changeable if required.

Standard frame used in partitions of different thickness

The 'snap-in' hinge consists of a steel male blade which is screwed to the frame. The female blade is smaller, made from reinforced nylon, and is screwed to the door edge. To remove the door the flexible snap piece on the bottom hinge is raised first and the male blade is withdrawn. The same process is then carried out on the top hinge. When replacing the door the top blade is inserted first and bottom blade second, the door is then snapped into position.

Mortise locks or latches are fitted. They are placed exactly central in the height of the door to enable inter-changeability. To complete the sets, architraves are also supplied. These, like everything else on the door set, are of the simplest design and two types are made. One is of plain section and the other is rebated to take up differences which may exist between thickness of partition and thickness of door frame. For speed in production architraves are not mitred at the angles, but are square cut. They are cut exactly to size in the factory and sent to site tied in bundles.

Fixing

As we have seen, door sets should be fixed after the plastering work has dried out, which is at the second fixing stage of on-site carpentry work. The following sequence of operations should be followed:

1   Remove profiles from opening.
2   Remove door from its frame.
3   Using wedges, place the frame into the opening, making certain that it is plumbed and squared accurately.
4   In the position of the fixing holes in the frame (made by manufacturer) drill pilot holes into partition walls.

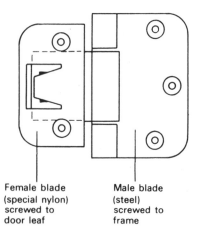

Female blade (special nylon) screwed to door leaf

Male blade (steel) screwed to frame

Snap-in hinge used on door leaf

5   Enter screws and insert packing immediately behind to prevent distortion when screws are tightened.
6   Remove wedges and snap the door on to hinges and make any final adjustment to ensure equal clearance all round.
7   Fix architraves.

Note: Particular care must be exercised in stacking door sets on site before use. This will ensure that they are perfectly flat when required for fixing.

Fire check doors

Although it is impossible to construct a door which will totally resist fire, in certain circumstances it is now necessary to make doors which will resist fire for a period of time. It is hoped that this will allow time for a building to be evacuated when fire breaks out.

Approved Document B of the Building Regulations relating to fire precaution work lays down two standards of resistance in door construction.

One hour fire resistance: this is required where the risk is at its greatest. These doors have a minimum thickness of 54 mm.

Half hour fire resistance: for situations which are less vulnerable. Doors have a minimum thickness of 44 mm.

Fire precaution works are required in public buildings, hotels and houses which are separated into flats with more than one tenancy. The materials used in the manufacture of fire check doors are materials which themselves are fire resistant, such as asbestolux and plaster-board. However, some very dense hardwoods are considered adequate, provided the whole door is of solid construction and of the required minimum thickness. These doors are not unlike flush doors in that they have an inner frame which is clad in approved materials.

Half hour fire resistance (FR) doors consist of the frame with plaster-board of 12 mm minimum thickness applied to both sides. For neatness this can be rebated in. The outer skin of plywood can then be fixed. Edges are lipped as in the case of ordinary flush doors. To increase the fire resistance of the door to one hour the construction is the same, with the addition of 6 mm asbestolux to cover both sides of the door immediately beneath the face plywood panel.

For one hour fire resistance all doors must be hung into solid frames (not linings) with rebate depth (for door stop) of at least 25 mm. In the case of half hour resistance, however, the fire officer may, in certain circumstances, allow linings, but the door stop thickness must be built up to 25 mm. Door sets are also manufactured with a maximum of a half hour fire rating.

All fire check doors must be made self closing by means of an approved door closer and *must not be wedged in the open position*. The joint around FR doors must not exceed 3 mm.

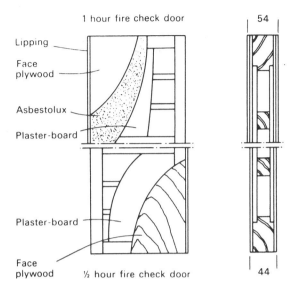

Fire check doors and (on the right) a section through door (enlarged) showing construction

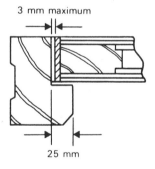

One hour fire resistance door and frame

The 'setting-out' rod

Before any item of joinery is commenced the work should be set-out on what is called a *rod*. This is particularly so if more than one item of the same size or design is required.

The rod is a board upon which the horizontal and vertical sections of the work are drawn full

size. There must be no broken lines or scaled sizes on the rod. The size of the rod will depend on the size of the work being set-out. In some cases wide softwood boards of about 12–15 mm thickness will be quite adequate. Plywood is also used extensively, but this can be expensive. A method now used a good deal is

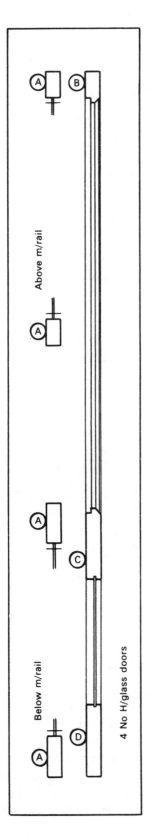

A typical rod

that of using decorators' lining paper placed on plywood as a backing and secured with masking tape. The work can then be set-out on the lining paper and removed when required. This makes the storage of rods much easier, especially if they are to be used again in the future.

All rods must have a planed straight edge which can be used to square from and act as a face edge. Only the essential information to enable the work to be done should be included on the rod. It is bad practice to include unnecessary information because this can lead to confusion and inaccuracy. Where a moulding of the same size occurs in a number of different positions it is quite sufficient to draw the moulding accurately only once. The machinist can make the cutter to the correct shape from this single drawing.

Once the work has been fully and accurately drawn on the rod a '*cutting list*' of materials required to complete the job can be prepared. The list has no special form, but it must contain a description of the item as well as all sizes and the type of timber from which it is to be cut. For cost effective reasons some firms include the sawn sizes of the material as well as the finished sizes on the cutting list. However, this is optional because most machinists will cut as little waste as possible from their timber stocks.

When all the timber has been prepared the various members are placed on the rod and accurately marked-off. These should be clearly marked 'PATTERN' and used to convey the setting out-markings to the other material only. By this means maximum accuracy should be maintained. If the item is to be repeated at a later date the patterns can be retained for that occasion, but otherwise the patterns can be machined, fitted and assembled at the end of the batch.

In the example illustrated it is assumed that four half glazed external doors are required. Because the stiles are diminished it means that there is a different detail below the middle rail than that above. Accordingly, this has to be shown on the rod.

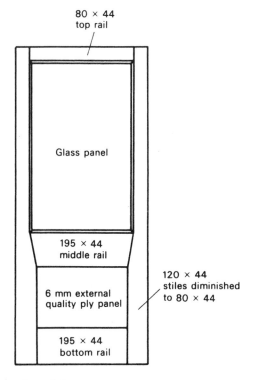

Elevation of door

Customers name:		J.FENN						
Description of job:		4 N° H/GLAZED EXTERNAL DOORS						
Item	Description	No. reqd.	Length	Sawn		Finished		Material
				W	Th	W	Th	
'A'	STILES	8	2.000 m	125	50	120	44	REDWOOD
'B'	T/RAIL	4	850mm	86	''	80	''	''
'C'	M/RAIL	4	850 mm	200	''	195	''	''
'D'	B/RAIL	4	850 mm	''	''	''	''	''

The cutting list

Questions

1 Using a sketch, show all members incorporated in a typical four panel door and name all the parts.

2 Describe the construction of a typical flush door.

3 With the aid of sketches, describe the principle of draw boring in frame construction.

4 Describe the arrangement of bracing door frames or linings for delivery to site.

5 What is the purpose of making the opening for a door lining larger than the overall size of the lining?

6 What method is used to form openings in walls or partitions when pre-assembled door sets are to be used?

7 Draw a vertical section through (a) a one hour fire resistant door and (b) a half hour fire resistant door, clearly showing the construction of each.

8 Describe the construction of a frame suitable to comply with one hour fire resistance.

7 Windows

The two main functions of a window are:

1 To allow natural light into the building.
2 To ensure adequate ventilation in the room. The Building Regulations state that the total ventilation area, that is the area of all opening sashes added together, must be at least one twentieth of the total floor area for all habitable rooms (rooms which are lived in).

Many forms of windows are in use at the present time, the following being among the most common:

1 *Casement window*
2 *Sliding sash window* (also known as box frames)
3 *Bay window*
4 *Metal casement window*

All windows are constructed with one or more glass openings known as *lights*. Bearing in mind that windows are subjected to all weather conditions they must be made in such a way as to provide maximum protection against the worst weather conditions.

Casement window

This type of window consists of a solid outer frame with one or more smaller and lighter frames, called sashes or casements, within it. The number of sashes included usually depends on the overall size of the frame. Those sashes which are hinged are called *casements*, while those which are not are known as *fixed sashes*. Outer frames are made up of a head, two jambs and a sill. When the height is divided it is done by introducing a *transom rail* which runs horizontally across the frame and tenons into the jambs. The width of the frame is divided by vertical members called *mullions*. The mullions must be tenoned into the sill and the underside of the transom. In many cases the mullion will continue above the transom by being tenoned into the top side of the transom and the head of the frame. The sashes above the transom rail are called *fanlight sashes*.

All joints used on a casement frame are of the normal mortise and tenon type. It should be noted that the head and sill contain the mortises while the jambs are tenoned.

To give better protection to the transom joint with the jamb, the over-run of the transom

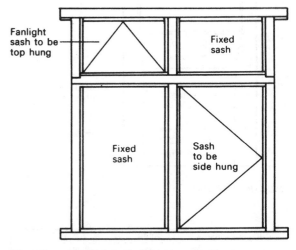

Elevation of window (without sashes)

should be let into the face of the jamb by about 6 mm.

The outer face of the frame is rebated to receive the sashes. The rebates should not be less than 12 mm deep. Chamfers or simple mouldings are applied to the inner edges.

Sashes are much lighter in construction than

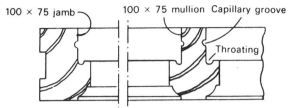

Part horizontal section

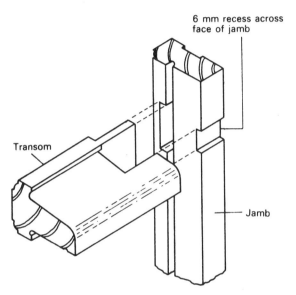

Detail of joint between transom and jamb

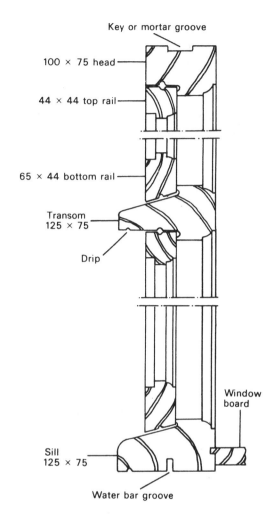

Vertical section through a four light casement window

frames. The two outside vertical members are called stiles. The upper rail is called the *top rail* and the lower is the *bottom rail*. In sash construction the rails are tenoned and the stiles mortised. They are constructed with mortise and tenons, but here we see a practice which is peculiar to sash construction. The tenons cannot be haunched because it would cut away too much of the sash stile, so a method known as *franking* is used.

A *franked mortise and tenon* is the opposite to the haunch. Whereas the haunch is left on and chopped into the upper end of the mortise, the franked portion is cut away to fork over the piece which has to be retained on the stile.

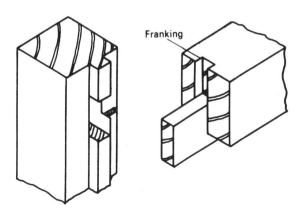

Sash corner joint showing method of franking

Sashes are rebated to receive the glass. These rebates should not be less than 9 mm deep and should always be on the outside of the frame.

Stormproof windows

As the name suggests, these windows have been designed to increase their performance against bad weather. Stormproof casement windows are a slight variation on the traditional casement. The solid outer frame remains basically the same, but the sashes are rebated and splayed all round the edge to allow the rebate to mask the joint, thereby reducing the possibility of driving rain etc. from entering the building.

Although the outer frame joints are normally mortised and tenoned the sash joints are framed with the comb joint. Therefore, most of this work is carried out by machine, leaving only the final assembly to be done by hand.

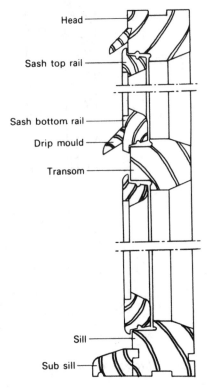

Vertical section through standard stormproof window

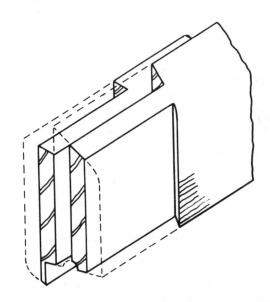

The comb joint applied to corner of stormproof sash

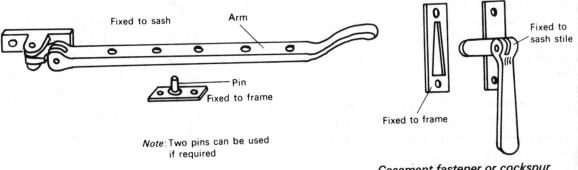

Casement stay

Casement fastener or cockspur

Casement ironmongery

Casement *stays* and casement *fasteners* are used to retain hung sashes in either the closed or partly open position. Casement stays consist of an arm containing a series of holes which locate the pin screwed to the frame and will hold the sash in the required position. As a locking device for fanlights a stay is used with two ordinary pins fixed apart, or a patent locking pin can be obtained.

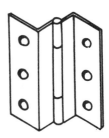

Cranked hinge for use with stormproof sashes

Casement fasteners are sometimes called *cockspurs* and are screwed to the sash stile. The tongue of the fastener enters a mortise on the frame which is covered by a slotted plate. The narrowest part of the slot must be at the top. Ordinary *butt hinges* are used for hanging sashes on normal casements, but a *cranked hinge* is used on stormproof sashes.

Sliding sash windows and box frames

Although these windows are of much older design than casements, they are still widely used.

The frame, which in many cases is called a box frame, surrounds sashes which slide up and down in a vertical movement. Methods used to bring about this vertical movement are:

1 *Cords and weights*: The sashes are attached to cords on the other end of which is a *sash weight* to balance the movement.
2 *Spring tape balance*: A simple coiled spring which is fixed to the frame and attached to the sash.
3 *Spiral sash balances*: A face fixed telescopic spring running inside an outer case.

The frames for this type of window are constructed almost entirely with the use of housing joints. Solid timber sills are used, but the head and vertical members are made up of pulley stiles, inside and outside linings, and back linings. In those windows where cords and sash weights are used the frame must be of the 'box' type of construction. This means that the head and jambs must be boxed, but the sill can be solid.

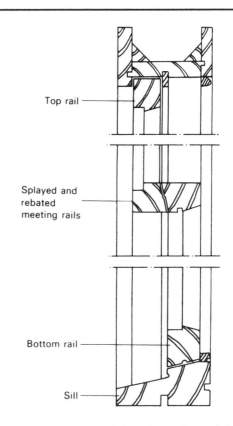

Top rail

Splayed and rebated meeting rails

Bottom rail

Sill

Vertical section through box frame for weighted sashes

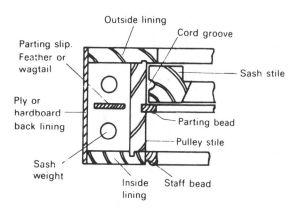

Part horizontal section

The *pulley stiles* and the *pulley soffit* (underside of head) form the inner faces of the frame, while the *inside linings* and *outside linings* make up the faces. The box is completed on the sides only by a *back lining* which is made up of ply, hardboard or timber off-cuts. The back lining is permanently fixed to fully enclose the box which houses the weights. The weights serving the upper and lower sashes are separated by a *parting slip* which is also sometimes referred to as the *feather* or *wagtail*. Although this must be thin enough to move easily when getting access to the weights it must not be so thin that it will snap and cause problems. A thickness of 6 mm is recommended.

Parting beads are grooved into the pulley stiles to separate the upper and lower sashes. *Staff beads* are applied on the inside of the window to allow the removal, when necessary, of the sashes for maintenance. Notice that the head of the frame is left open and glued angle blocks are used to give support to the inner and outer head linings.

Before assembling the frame, provision must be made for the removal, when required, of the sash weights. This is done by forming *sash weight pockets* on the inside lower portion of the pulley stile. Make certain that the splayed cut is at the higher position to prevent the pocket

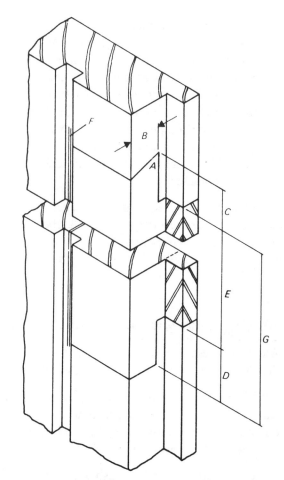

A = 30° splay cut at top end of pocket
B = Half thickness of pulley stile
C = 12 mm approx overlap of cuts (upper)
D = 15 mm approx overlap of cuts (lower)
E = Portion of tongue to be removed
F = Saw cut midway of parting bead groove and full length of pocket
G = Overall length of pocket varies according to length of weight

Sash weight pocket detail

from falling forward. The illustration shows the method of setting out and cutting the pocket.

Sash weights are made of cast iron. They are either round in shape or square. The sash must be weighed fully glazed and the total weight of the two weights attached must equal that weight. In practice it will be found more effective when the weights of the upper sash

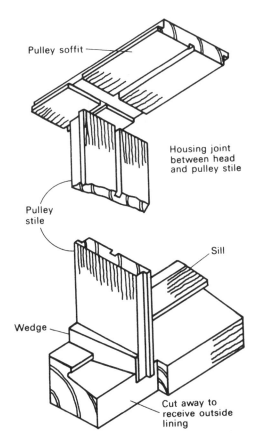

Pulley soffit

Housing joint
between head
and pulley stile

Pulley
stile

Sill

Wedge

Cut away to
receive outside
lining

Method of securing pulley stile in sill housing

together weigh 250 g more than the sash (when fully glazed) and the weights for the lower sash are 250 g less than the sash. This will ensure that the upper sash is kept up by the slightly heavier weights while the lower sash, with lighter weights, remains firmly in the lower position — producing the best possible joint at the meeting rails.

The *axle pulleys* over which the cord travels between the sash and the weights should also be fitted prior to assembly.

A wedge is cut in behind the portion of pulley stile which is housed into the sill. Assembly should start by gluing these wedges and securing the pulley stiles. Where methods other than weights are used for balancing the sashes, the outer frame need not be of boxed construction. However, it will be appreciated that thicker pulley stiles will be required because the spring tapes or spiral balances will be directly fixed to them. Staff beads and outside lining pieces will still be required to contain the sashes.

Sliding sashes are constructed to make them capable of standing up to the heavy usage placed upon them. The meeting rail on the

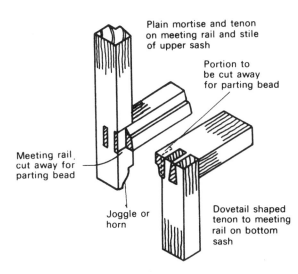

Plain mortise and tenon
on meeting rail and stile
of upper sash

Portion to
be cut away
for parting bead

Meeting rail
cut away for
parting bead

Joggle or
horn

Dovetail shaped
tenon to meeting
rail on bottom
sash

Detail of meeting rail joints

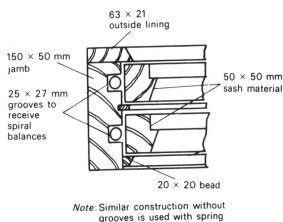

63 × 21
outside lining

150 × 50 mm
jamb

25 × 27 mm
grooves to
receive
spiral
balances

50 × 50 mm
sash material

20 × 20 bead

Note: Similar construction without
grooves is used with spring
tape balance

Alternative frame construction when spiral sash balance is used

upper and lower sashes must be fitted well if wind and rain are to be kept at bay. The upper sash has a top rail and two stiles to complete the frame while the lower sash has a bottom rail and two stiles. Franked mortises and tenons are used on the joints between stiles and top and bottom rails. However, the joints between stiles and meeting rails should be studied closely. The meeting rails are wider in order to take up the gap between the upper and lower sashes created by the parting bead. This extra width is splayed to form a weather-tight joint. A small rebate is sometimes included to further improve weather resistance and reduce the possibility of illegal entry by sliding a thin piece of metal inside the splay and releasing the catch.

It is usual to form a *joggle* on the lower end of the upper sash stiles. This is rather like an elongated horn which can be shaped or left plain. The joggle enables a through mortise and tenon to be used on the upper sash between the stile and meeting rail.

Where no joggle occurs on the lower sash a dovetail shaped tenon must be formed so that the rail will not pull out when the meeting rail is used to open the window. The overhanging part of the meeting rail should be recessed into the face of the stile by about 4 mm. This avoids forming a sharp edge which may allow dampness to attack the joint.

Glazing bars are sometimes used to form smaller panes of glass in large sashes. These are usually stub tenoned into stiles or rails, but greater strength will be achieved at the intersection of vertical and lay bars if a halving joint such as that illustrated is used.

Sliding sash ironmongery
The sashes must be weighed fully glazed before the correct weights can be obtained. The cords to which they are attached are made from various materials, the most common being best

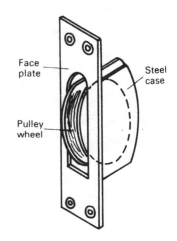

Standard pattern axle pulley (brass wheel and face plate)

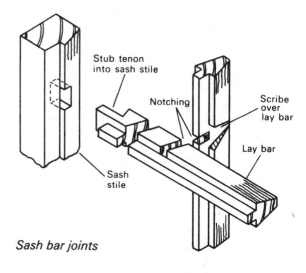

Sash bar joints

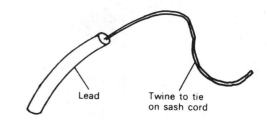

A mouse

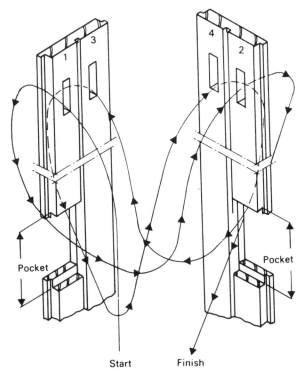

Method of cording sashes in a boxed frame

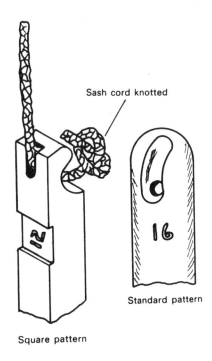

Types of sash weight with weight (in pounds) marked

twine or nylon. Cords vary in thickness from 5 mm to 12 mm according to the sash weight, the larger size being for the much heavier sash. Cords must be nailed into the groove in the sash stile with galvanised clout-head nails. The nails used are galvanised to prevent rusting and clout-headed because this is a large round head which is unlikely to pull through the cord.

Cording is carried out with a *mouse*. Most craftsmen make their own mouse with a piece of lead approximately 50 mm long hammered around a length of strong twine until it is roughly 7 – 9 mm in diameter. The twine is tied to the end of the sash cord and the mouse is then inserted into the box frame through the axle pulley wheel. The weight of the lead drops to the level of the open pocket in the pulley

stile and can be pulled through quite easily bringing the sash cord with it.

A very efficient cording sequence is as follows:

1 Thread the cord using the mouse over the *left-hand inside* axle pulley.
2 Bring the cord through the pocket and into the *right-hand inside* axle pulley.
3 Take the cord from the pocket to the *left-hand outside* axle pulley.
4 Finally, the cord is threaded into the *right-hand outside* axle pulley.

The weight should be attached to the cord by inserting it into the hole provided and tying a large knot. This is done with each cord in the reverse order to which they were threaded.

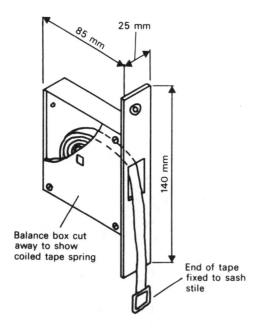

Spring tape sash balance

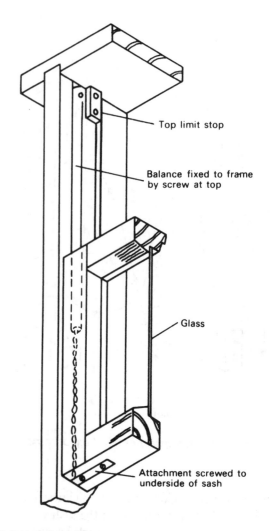

Spiral sash balance

Spring tape balances

Spring tape balances must also be of the correct type to suit the sash weight. The spring tape is contained in a metal case which is fitted into the pulley stiles at high level while the free end of the tape is attached to the edge of the sash stile. Two spring tape balances are required for both upper and lower sashes.

Spiral sash balances

Spiral sash balances are reasonably simple to install. Various types are manufactured to suit the different weights of sash. A rod of flat metal which is twisted rather like a corkscrew and called a *spiral rod* is fixed to the sash. The rod is inserted into a tubular metal case of varying diameter. A spring inside the tube retains the sash in required positions. The tubular cases become larger as the sashes become larger and heavier. The case can be grooved into the stile of the sash or pulley stile, but the groove must travel the full length of the sash run. The only means of fixing is one screw at the top of the

tubular case. After fixing the case, the spiral rod is threaded into it until the correct tension is attained. The attachment fitting is then screwed to the underside of the sash. An additional channel fitting is supplied for those sashes with joggles. Finally, limit stops are fitted to prevent over-run of the spiral rod. A short stop is used for the top of the window and a longer one for the bottom.

Having fitted the sashes and obtained a good sliding balance a brass or steel *sash fastener* is fixed on the upper flat surface of the meeting rails. The part of the fastener containing the lever is screwed to the lower sash nearest to inside, while the socket section is screwed to the upper sash.

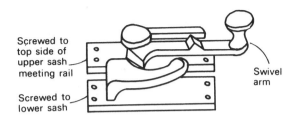

The fasteners are designed to pull the upper and lower meeting rails together. When properly fitted, this creates a weather-tight finish.

Sliding sash fastener

Sash thumbscrews are sometimes fitted for extra security. These are also made in brass or steel and are fitted through the inner meeting rail into a threaded plate in the outer meeting rail.

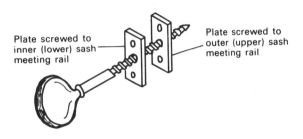

Sash thumbscrew

Bay windows

Windows which project from the face of a building are called bay windows. The various types of bay window take their name from the shape they form when seen in plan view. The most common being:

Square: This type has return sides which are at 90° to the front portion.
Cant or splayed: In this type the return sides are at an angle greater than 90°
Segmental: Partly circular, but not a full semi-circle.

Most bay windows are made in a similar way to the outer frame of the casement window. However, in the segmental type *angle posts* are added to the corner of the front and returns. These angle posts can be built up with two pieces of timber or made out of the solid. Either way, the angle posts must be tenoned into the sill at the bottom and the head at the top in the same way as the jambs.

The heads, sills and transoms (where used) are mitred at the angles and for additional strength either *hardwood dowels* or *handrail bolts*, or both, are used to secure the strength of the joints. It will be noted that although the heads, sills and transom (where used) are shaped in the segmental bay the sashes are not, this avoids the need to use very costly circular panes of glass.

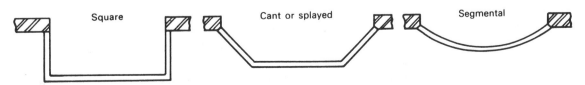

Common plan forms of bay windows

Quite often the sashes in bay windows are, like casement sashes, made from timber, but it is also common to find metal casement sashes incorporated into timber outer frames. As with casement windows, some are fixed lights while others are made to open for ventilation.

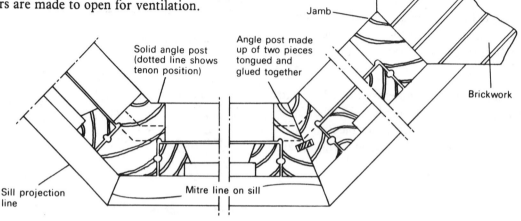

Alternative angle posts

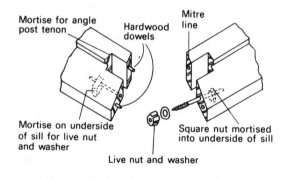

Method of jointing sills, transoms and heads at mitre

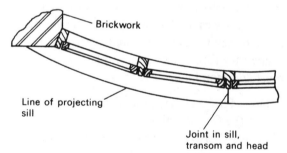

Segmental bay with flat sashes

Metal casements

Metal windows were introduced in the belief that they would be much more durable than timber ones. This will only be true if they are properly and regularly maintained. Corrosion and other problems can reduce the window to an unsatisfactory condition. Manufacturers have produced a very wide range of modular sizes. This means that it is possible to bolt small and large lights together to suit any particular requirement. It also makes replacement of such windows an easier task.

In factory and warehouse construction metal casements are often built into the brickwork with no timber outer frame. However, a timber surround is more commonly used nowadays. Because it has no structural function the outer frame need not be as heavy as those used for casements, but it should be mortised and tenoned for strength. The sill of the surround should be hardwood to afford greater resistance to water or dirt which is likely to collect on it.

The head, jambs and sill should be double rebated to receive the metal frame, which is bedded in waterproof mastic and screwed to the timber surround. Accuracy in the formation of the double rebate is important to ensure that a water-tight joint is formed between the metal and timber. Timber transoms and mullions are not used in these frames, being replaced by metal ones of lighter design.

Key grooves should be formed all round the outer edge of timber frames which are to be built in. These are sometimes called *mortar grooves* because they allow the cement mortar to go into the groove, thus forming a key and giving additional strength to the fixing.

Water bars are also used on the underside of the sill to prevent water or dampness from

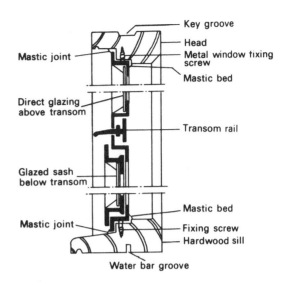

Section through metal casements in timber surround

penetrating at this point. The bar is normally about 25 × 6 mm in size and made of galvanised mild steel. Ironmongery used on metal casements is similar to that on timber casements, but instead of being screwed on by the carpenter it is either welded or bolted on before leaving the factory.

Double glazing

Windows are glazed with glass of varying thickness and pattern. Most window glazing is done with putty, but the glass may also be secured by beads which are pinned on. Beaded glass must always be bedded in putty.

Glass is a poor insulator and considerable heat loss is caused by single glazing. For this reason *double glazing* is now widely practised,

especially in urban areas where it also helps to keep out traffic noise etc. Double glazing is known to cut down heat loss by at least 50 per cent. In days gone by it was the practice to construct frames which had two sashes fitted (one inner and one outer). Nowadays, however, it is much more common to find double glazing units which are made up and ready for use.

Ready-made units consist of two glass sheets with a space between them in which there is *dehydrated air*. The space can be 5, 6, 9 or 12 mm and the greater the space the more efficient the unit. Where these units are to be used, the rebates have to be made large enough to accommodate them. Where possible, double glazing units should be beaded, especially if they are of the thicker variety.

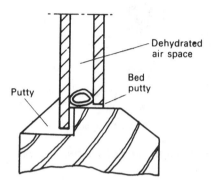 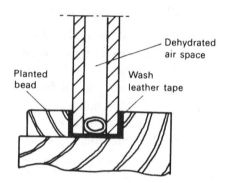

Typical uses of double glazed sealed units

Questions

1 Sketch a six light casement window and name all members.

2 Where would the following be found and what are their function:
(a) drip,
(b) mortar groove,
(c) throating.

3 What is the reason for franking the joints on a sash corner?

4 Why should the narrowest part of a casement fastener mortise plate be at the top when fixed?

5 Give reasons why it is useful practice to increase the upper sash weights and decrease the lower sash weights on a pair of sliding sashes.

6 Sketch the section of sliding sash meeting rails which will:
(a) give maximum security against unlawful entry and
(b) provide better weather protection.

7 What advantages can be obtained from double glazing?

8 Timber floors and flooring

Timber floors are constructed to suit a variety of different situations. The type of construction will usually depend on the position of the floor, i.e. ground or upper and the size of the span. In this chapter we shall be looking specifically at two types of timber floor:

1 *Suspended timber ground floor*: The type found on the ground floor of most detached and semi-detached properties. There is a space between the floor and the over-site concrete.

2 *Single or upper floor*: Found in most dwellings at first floor level when the floor is of timber construction.

Suspended timber ground floors

The Building Regulations, among other things, set out to prevent damage to buildings by either moisture or fire. Firstly, to comply with the regulations all such floors must be so constructed as to prevent the passage of moisture from the ground to the upper surface of the floor.

1 A *damp proof course* (DPC) must be inserted at not less than 150 mm above ground level. DPC's are a water proof membrane inserted into the cement mortar course between the brickwork. Several types are in use, the most common being pitch polymer or a lead core sandwiched in layers of felt. They are bought in rolls of about 20 m length and conform to brick widths, e.g. 112 or 225 mm.

2 No timbers are allowed to be built in below the DPC. Therefore, any moisture creeping up through the brickwork would be halted at the DPC before it reaches any timber.

3 If a fireplace is to be built a concrete hearth of not less than 125 mm thickness must be provided. The hearth should extend not less than 500 mm in front of the chimney breast and must extend beyond the inner

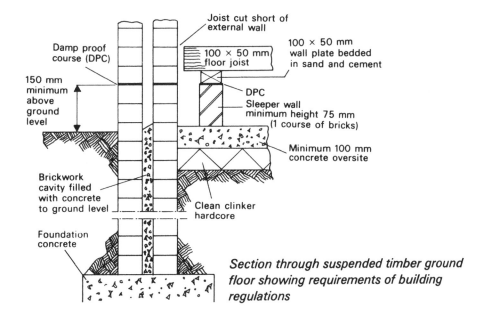

Section through suspended timber ground floor showing requirements of building regulations

face of the brickwork jamb, not less than 150 mm on each side. This means that no timber must be closer than these measurements to the fire opening. Thus, the risks of fires being started by falling embers is reduced.

Air-bricks are built into the external walls of the building to allow a continual flow of air to ventilate the under-floor area. This prevents situations of dampness without ventilation, which can quickly lead to dry rot or other forms of decay. Sawn softwoods, preferably treated with an approved preservative, are the best material for these floors.

Under-floor preparation

Before any carpentry work can commence a 100 mm concrete sub-floor must be laid on a bed of hardcore made up of clean clinker. The upper surface of the concrete must not be below ground level. It is on this concrete layer that the sleeper walls are laid in honeycomb brickwork. Honeycombing consists of laying the bricks in such a way that they have gaps between them in each course to allow a free flow of air throughout the under-floor area.

In timber ground floors at least one course of bricks surmounted by a wall plate must be laid to retain the air circulation. Sleeper walls are built up to DPC level and the 100 × 50 sawn timber *wallplate* is bedded down and levelled in on sand and cement mortar to provide a fixing for the floor joists.

The first sleeper wall is built near to, but not touching the external brickwall, and others are built at approximately 1.8 m centre to centre finishing with a sleeper wall close to the other wall. Timber should not come into contact with the external walls at any point.

The under floor wall which supports a hearth slab must be in solid brickwork and is called the *fender wall*.

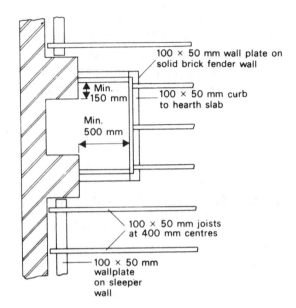

100 × 50 mm wall plate on solid brick fender wall

100 × 50 mm curb to hearth slab

Min. 150 mm

Min. 500 mm

100 × 50 mm joists at 400 mm centres

100 × 50 mm wallplate on sleeper wall

Typical lay-out of joists around fireplace at ground floor level

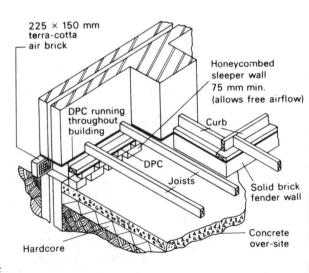

225 × 150 mm terra-cotta air brick

Honeycombed sleeper wall 75 mm min. (allows free airflow)

DPC running throughout building

Curb

DPC

Joists

Solid brick fender wall

Hardcore

Concrete over-site

Isometric view showing ventilation of underfloor

Joints

When all plates have been levelled-in and bedded down the joist arrangement can commence. Although the plates are not nailed in any way the joists should be nailed to the plates, taking care that the top edges all line through accurately because it is on this plane that the flooring will be laid.

A fairly long straight edge should be used for this purpose and where discrepancies occur packing pieces can be used to restore accuracy.

Joists are usually 100 × 50 mm sawn softwood and are fixed at approximately 400 mm centres making certain that there is a gap between every joist end and the wall adjacent to it.

During the construction of this type of floor the carpenter should be constantly checking that no timber is likely to be adversely affected by moisture or dampness. Equally, it is very important to leave the whole under-floor area clean and free of all material or dust which is likely to attract moisture.

Single or upper floors

This type of floor represents a different approach in constructional methods from the timber ground floor. The joists have to span from wall to wall with no intermediate support. This is why the joists are called *bridging joists*. Where fireplaces or stair openings penetrate the overall floor area the openings have to be trimmed. This must be done correctly if a strong rigid floor is to be constructed.

The underside of the upper floor is the ceiling

of the room below, so it is necessary for accuracy to be maintained on both sides to achieve best results. The Building Regulations stipulate that the clear vertical height from floor to ceiling must not be less than 2.3 m for habitable rooms. This does not include space used purely for storage purposes.

Joists

Bridging joists are normally sawn softwood 50 mm thick while those joists which are used

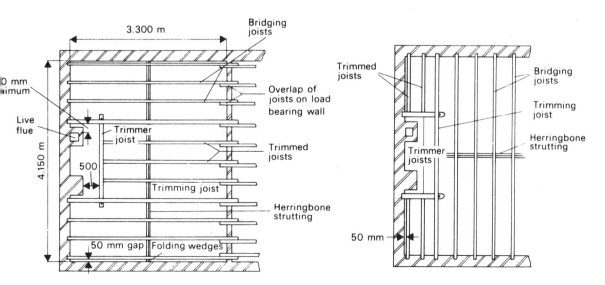

Alternative methods of upper floor joist lay-out

to trim openings are called *trimmer and trimming joists*. Because of the loading placed upon them these joists are half as thick again as the bridging joists, being usually 75 mm thick. The depth of the joists will normally be calculated at the design stage of the building, but where there is any doubt a useful 'rule-of-thumb' method of determining the depth is to divide the span (in *decimetres*) by 2 and add 2. This will provide the joist depth in centimetres.

Example: Span = 4.000 m

$$\text{Therefore, depth} = \frac{40}{2} + 2$$

$$= 22.$$

The joist depth, therefore, is 22 cm or 220 mm. As 225 is a standard size that is the size that would be used in a span of 4.000 m to avoid waste and to save time.

To comply with the Building Regulations a hearth slab of 125 mm concrete would have to be constructed when a fireplace is built. This

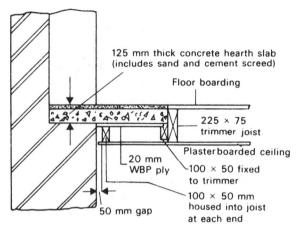

Section through hearth showing formwork of slab

would involve the carpenter in forming the shuttering for the concrete to be poured. However, this shuttering is often left in position because it cannot be seen when the ceiling is plastered. The minimum size of hearth for upper floors is exactly the same as for under-ground floors.

Joists

The joists in a single or upper floor can run in any direction, but as a general rule they are laid across the shortest distance. This is not only results in a more rigid structure, but it can also prove to be more economical with less waste of material.

Joists are supported on each end in any of the following ways:

Direct bearing
In this method the inner thickness of brickwork is built up to the underside of the joist. The joists are then laid directly on to the brickwork. *Note*: This method can cause difficulties due to the irregular sizes of bricks giving an uneven line of ceiling.

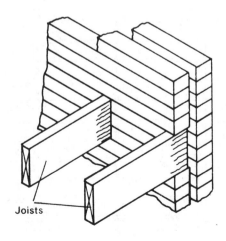

Joists bearing directly on brickwork

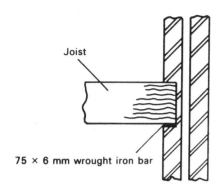

Wrought iron bar bearing

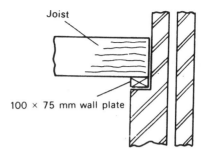

Built-in wall plate

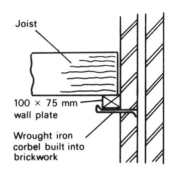

Wrought iron corbel and wall plate

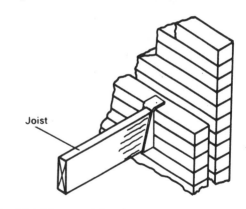

Steel joist hanger (shoe)

Wrought iron bar

This is similar to the direct bearing method except that a wrought iron bar 75 × 6 mm thick is laid on the brickwork. The bar has the effect of keeping the underside of joists reasonably accurate and also distributes the loading placed on the joists.

Built-in wall plate

Here, the brickwork is built up and a sawn timber plate 75 × 100 mm is bedded on top. Joists are nailed to the wallplate.

Corbel and wall plate

Wrought-iron corbels (brackets) are built into the brickwork in the required positions. The wall plate is then seated on the corbel. This method leaves the plate visible from the room below.

Joist hangers

Steel joist hangers (sometimes called 'shoes') are approved joist supports by all authorities. They are used in the process of floor trimming as well as where the joist end abuts the wall. With this method no part of the joists need to be built in, thereby reducing the possibility of decay.

When joist hangers are not used and floor trimming is required a joint called the *tusk tenon* is used. If the correct setting-out is followed the tusk tenon will provide a joint of immense strength. Apart from the inclusion of the tusk to give strength to the joint it will be noted that the underside of the tenon is on the

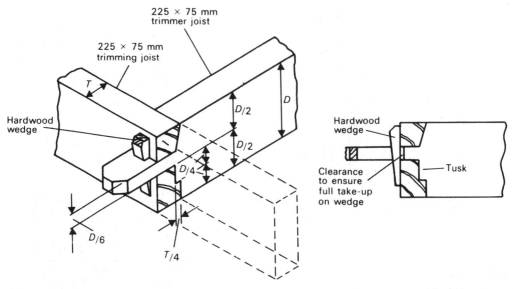

Tusk tenon proportions

centre line of the joist depth. This means that the tenon is situated entirely in the upper half of the beam — the half which is in compression — whereas the lower half is in tension when the joist is loaded from above. If the tenon was subject to this tension it would be much more likely to break down.

The joist which bears the tenon is called the *trimmer joist* while the joist with the mortise is

called the *trimming joist*. Any joist which has to be cut and housed into the trimmer is called a *trimmed joist*.

Single or upper floor joists, like ground floor joists, are spaced at approximately 400 mm centre to centre. If after measuring the building it is found that the spacing has to be slightly under or over, it is quite permissible, always provided that it is not by too great a margin.

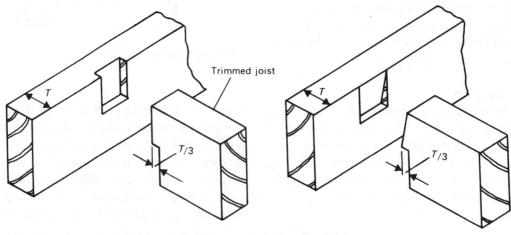

Methods of housing trimmed joists into trimmers and trimming joists

A further requirement of the Building Regulations in single floor construction is that no timber should be built in with its face nearer than 200 mm to the inside face of a live flue, that is a flue serving a solid fuel fire.

Strutting and bridging

It will be understood, from our knowledge of timber characteristics, that the longer a piece of wood is the more likely it is to distort. This is true with floor joists. Therefore, when a joist is more than 3.5 m in length a row of strutting or bridging is fixed midway in its length and between each joist. This holds them rigid and prevents movement or distortion. Folding wedges are placed between the brickwork and the first and last joists to ensure that the stiffening runs unbroken from wall to wall. The bridging or strutting is carried out in one of the two following ways:

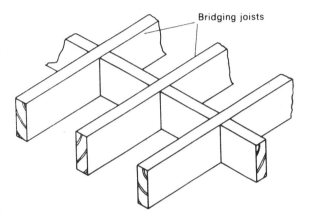

Solid bridging between joists to prevent movement

1   *Solid bridging or strutting*: In this method solid pieces of timber (very often the off-cuts from the joists) are cut tightly between each joist and skew nailed through the sides to hold them in position.
2   *Herring-bone strutting*: This is the most common method used. It consists of pairs of battens, approximately 32 × 50 mm, cut diagonally between the joists and nailed through the edge of the batten on the splayed cut into the joist. In practice it is wise to put a small saw cut in the pointed ends of each strut to prevent splitting when the struts are nailed.

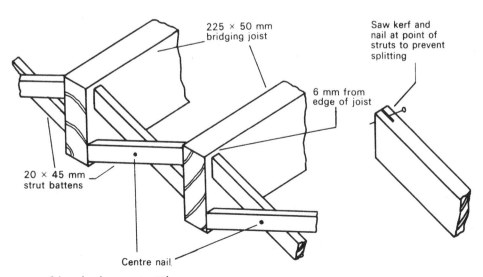

Arrangement of herringbone strutting

Insulation

One recurring problem with upper floor construction is the transmission of sound from the upper to the lower rooms, especially when the floor is walked upon. Architects have devised some quite elaborate methods of insulating single floors against sound, but cost does not always allow such measures to be taken. If a floor is carpeted the sound is deadened quite considerably. One fairly inexpensive method of reducing sound transmission is illustrated. It involves the fixing of 12 mm insulation board immediately beneath the flooring and also between the plaster boards and joists on the underside. Additionally, small

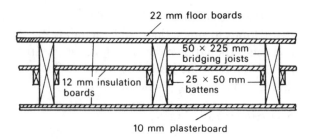

Section through upper floor showing sound insulation measures

battens are nailed about halfway down the joist sides and insulation boards are placed between each joist over the whole floor area.

Flooring

Tongued and grooved (T and G)

This boarding is used very widely as a floor covering at either ground or upper floor levels. The tongue is slightly off-centre on the edge of the board and the correct way for them to be laid is with the wider side of the tongue uppermost. This allows for the extra wear placed on the walking surface of the boards.

Plain tongued and grooved (note widest shoulder uppermost)

Splayed, tongued and grooved (for secret nailing)

Types of flooring in general use

Splayed tongued and grooved (ST and G)

These boards are similar to ordinary T and G, but have the tongue and groove formed in such a way as to allow them to be secretly nailed. The slight splay which is applied to the top of the tongue is designed to make nailing easier.

Square edge (SE)

Square edge boards are plain boards laid to abut each other. They are seldom used now because the shrinkage which inevitably takes place after laying produces cracks between the boards, thus leaving an unsatisfactory floor surface.

With the exception of secretly fixed boards, these types of flooring are nailed to the joists with a special nail called a *floor brad*. These nails are pressed out of sheet steel and are about 55 mm in length. The head is designed to embed itself into the board when the last hammer blow is struck. This makes punching-in unnecessary. It is much better to use either a lost head nail or an oval brad when fixing floor boards requiring secret fixing. These have to be fixed one at a time and therefore take much longer than other boards to complete the floor.

T and G or plain boards, on the other hand, can be laid in batches, provided they are properly cramped up with as tight a joint as possible before being finally nailed down. Patent floorboard cramps which attach to the joist are available to ensure that good tight joints are achieved, but in practice, a simple folding wedge arrangement is more often used.

A pair of folding wedges are cut from the material being used for the flooring. One is nailed temporarily to the joist and the other is driven on to the board's edge to form an effective cramping. As many pairs of these can be used as is felt necessary, bearing in mind the length of the boards and their shape. Fewer cramps will be necessary with flooring which is fairly straight.

Patent metal cramp being used to cramp boards

Floorboard joints

As it is not possible to lay all full length boards when covering a floor it becomes necessary to join shorter lengths. These joints must be made immediately over a joist and are called *heading joints*. When boards are jointed in this way the heading joints should be staggered over the whole area and not cut in close proximity to each other. The two most commonly used are:

1 *Square heading joint*
Both boards are cut square across the face, but should be slightly undercut to ensure that the joint on the face is as close as possible.

2 *Splayed heading joint*
A 45° cut is applied to the ends of both boards. This joint has the advantage of only having to be nailed singly.

Access to services

It is in the construction of floors perhaps more than any other stage of the building that the carpenter needs to work in close co-operation with other trades. Services such as water and electricity usually run beneath the floor,

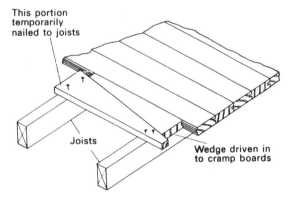

Cramping boards with folding wedges cut from flooring

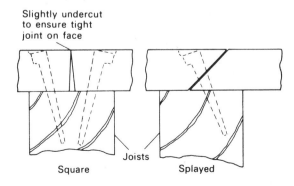

Types of floorboard heading joint

resulting in the need for joists to be notched-out or bored to accommodate piping or cables. Joists should not be cut away to such an extent that their structural strength is reduced or impaired. The maximum limits for holes and notches permitted by British Standard (B.S.) 8103 (Pt 3) are illustrated below. These should be strictly adhered to.

Small access points with easily removable covers can be formed in positions that are agreed between the trade supervisors. In most instances these covers must be screwed down and not nailed. Where long lengths of boarding need to be removable for access purposes the tongues (where they exist) must be removed. Such boards must also be secured by screws so that they can be withdrawn easily when required.

Chipboard flooring

Chipboard flooring is very widely used nowadays, but care must be taken to ensure that only flooring quality chipboard is used. Sheet sizes are 2440 × 600 × 18 mm. The joists should be spaced to take these sizes into account to avoid unnecessary cutting. The sheets are tongued and grooved on all four edges and it is quite common for chipboard sheets to be laid end to end, even where they do not have the support of a joist immediately beneath them.

Because of their greater size, chipboard flooring sheets can be used to cover a large area in a short time. This is one of the reasons why builders use them, especially on housing developments.

Another advantage is that they are in no way affected by woodrot or beetle attack. However, there is still the need to provide access to underfloor services and this usually means cutting panels in convenient positions. The panels are secured by screws so that they can be easily and quickly removed when required.

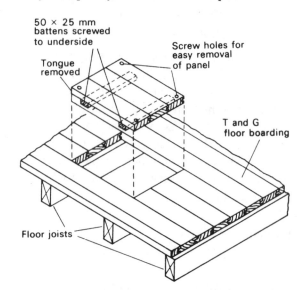

Method of forming small removable panel in floor

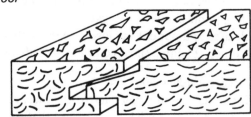

Chipboard flooring with tongued and grooved edges

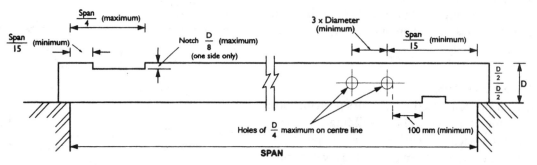

Notches and holes to satisfy B.S.8103 (Pt 3)

Questions

1 Described the function of a DPC.

2 What is the minimum number of brick courses permitted in a sleeper wall?

3 What is the function of an air brick?

4 Describe the possible consequences of inefficient underfloor ventilation.

5 Why are plates and joists positioned away from exterior walls?

6 What is the minimum vertical height permitted to the underside of an upper floor?

7 Describe the differences between:
(a) bridging joist,
(b) trimmer joist,
(c) trimming joist.

8 What is the function of herring-bone strutting and when is it necessary?

9 Sketch a tusk tenon and include the proportions for setting it out.

9 Roofs

Flat roofs of timber construction are not unlike the upper floors studied in the previous chapter. These will be looked at in greater detail towards the end of this chapter. Pitched roofs, however, vary quite considerably in size and type. For this reason the woodworker will need to study the principles involved when designing a pitched roof to suit a certain building. Of course, the size of the building is a very important factor in deciding what form of construction will be most suitable.

Roof pitch

The pitch of a roof is expressed in a number of different ways, the most common being:

1   In degrees, that is to say 30°, 45° or 60° etc.
2   It is sometimes expressed as a rise and span.

The *rise* being the vertical height measured in the centre of the span. For the purposes of roof construction the *span* is the horizontal distance over all the wall plates.

3   On more rare occasions it may be called one third or one half pitch. This means that the rise will measure one third or one half the span, respectively.

The size of the building affects the type of roof and as the span increases, so too must the complexity of construction to cope with the additional loads placed upon it. This rule does not apply to the simple *lean-to* roof which rarely extends to large spans. The two basic forms of pitched roof construction are:

1   Single roof construction
2   Double roof construction

Single roofs

The names given to some simple roofs and, in some cases, the maximum span for which they are suited are as follows:

Lean-to

This type is not often used on large spans, but more often found as a covering for a rear or side extension to a larger building. The lean-to varies to suit requirements and the upper end of the rafter is fixed to a *wall piece* which can be corbelled or bolted to the wall. The foot of the *rafter* is normally birds-mouthed (notched) over a *wall plate*.

Couple roof

This is a double pitched roof of the most simple kind, consisting of *common rafters* birds-mouthed over the wall plate and fixed to a *ridge board* at the apex. In general the birds-mouth

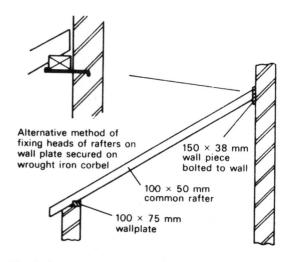

Alternative method of fixing heads of rafters on wall plate secured on wrought iron corbel

150 × 38 mm wall piece bolted to wall

100 × 50 mm common rafter

100 × 75 mm wallplate

Simple lean-to roof

notching should not cut into the rafter any more than one third of the depth of the rafter. The maximum span for this type of roof is 2.5 m.

Couple close roof

This type is suitable for spans up to 4 m. It is stronger than the couple roof, having a *ceiling joist* spanning the wall plates. This new member prevents the spread created by the forces distributed to the wall plates by the rafters, thus making it suitable for the larger span.

Collar-tie roof

This type of roof is suitable for spans greater than 4 m, but up to a mximum of 5.5 m. There is no specified position of the collar, but its effectiveness would be reduced considerably if it was placed any more than half-way between the wall plates and ridge. The most common position used is one third of the height of the rise above the plates. If the collar is halved into the common rafter with a dovetailed halving, as shown, the full strength of the tie will be achieved.

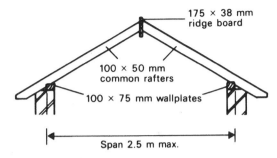

Couple roof

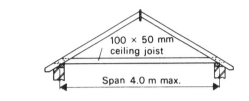

Couple close roof

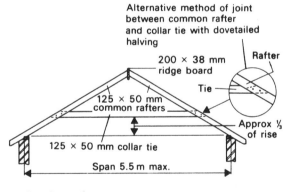

Collar tie roof

Double roofs

When the span of a roof is 6 m or over, *double roof construction* is used. This is because a secondary means of support for the roof becomes necessary and it is at this stage that purlins, binders and struts are brought in. In various forms a *purlin roof* can be suitable for spans up to 10 m. A common rafter spacing for

roofs of this type is 400 mm. *Purlins* are positioned on the underside of rafters about half way up the slope of the roof to give the structure support. *Binders* are fixed to the top side of the ceiling joists to 'bind' them together. This is because larger spans require greater lengths.

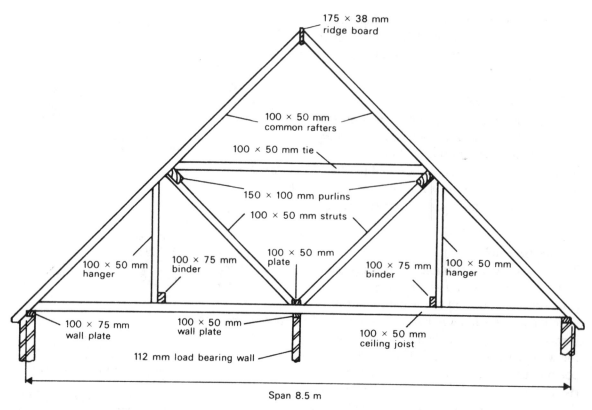

Purlin roof for large spans (double roof construction)

Ends of pitched roofs

Pitched roofs can be terminated at the flank walls of a building in two ways:

Gable ends

In this type the brickwork on the end wall of the building is carried straight up to form a triangle. Usually, the last pair of common rafters are fixed about 50 mm away from the flank wall brickwork which follows the same pitch line. The tile or slate battens are then fixed and cut off flush with the outside face of bricks. This allows the roof covering to be

finished with a small overhang to throw rainwater clear. This overhang is called the *verge*.

Gable ends are sometimes terminated with a *barge board*, which is more of a decorative feature than a functional one. In addition to the common rafters inside the flank wall another pair are fixed abutting the wall on the outside. This pair of rafters are secured by means of *noggings* fixed at about 600 mm centres nailed to both the inner and outer pairs of common

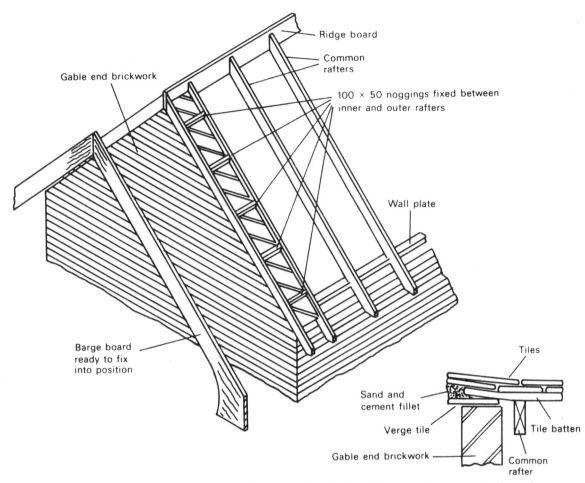

Gable end with barge boards showing gable ladder

Detail of gable end when verge is formed instead of barge board

rafters. This feature is also commonly called the *gable ladder* because of its appearance. It is the outside rafters which provide the base to which the barge board is fixed.

Barge boards are normally about 25 mm thick with varying widths from 175 mm. If desired the underside edge can be shaped as was customary many years ago.

Hipped ends

This type is created by continuing the slope, on the back and front of the roof, around the ends. In other words, the roof pitch occurs on all

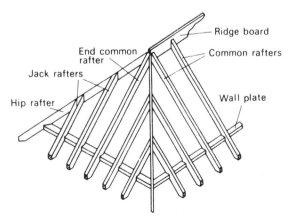

Hipped end of a pitched roof

sides of the building. This involves a great deal of extra carpentry work because not only are additional members of the roof required, but it also becomes necessary to accurately work out the bevels and true lengths of these additional roof members.

The main feature of the hipped end are the *hip rafters* which run upwards from the corners of the building to intersect with the ridge board. The hip rafter is always much deeper than the common rafter, often 225 mm, but is also thinner, some being as little as 25 mm, but more often 32 mm thick.

Jack rafters are cut in at the same spacings as the common rafters to infil the complete hipped end area. Although the birds-mouth on the jack rafters are identical to the common rafter, the upper end has a splay cut to fit against the side of the hip rafter. It will be noted that the jack rafters gradually *diminish* in length. These jack rafters are also widely known as *cripples*. This term is believed to derive from the fact that they are common rafters cut short, or 'crippled'. In the case of a purlin roof, the purlin must return across the hipped end. This also requires the development of angled cuts to ensure that an accurate fit is made at the angles.

Eaves detail

The portion at the foot of the rafters which overhangs the wall face is called the *eaves*. It is here that the slope of the roof terminates in a gutter, which is provided to collect rain etc. and carry it away to a rainwater downpipe and from there into a gulley.

Sprocket pieces are sometimes used at this end of the rafter to give a tighter joint to the slates or

tiles. They also ease rainwater into the guttering and prevent it from cascading over the edge. Sprockets are either fixed by nails on top of the rafter edge or, in some cases, nailed to the side.

Fascia boards of about 25 mm thickness are nailed to the front edge of the rafters to provide a back board for the gutter to be fixed to.

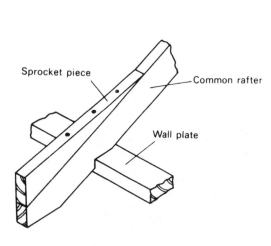

Sprocket piece nailed to top side of rafter

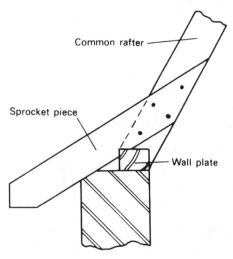

Sprocket piece nailed to side of rafter on a steep pitched roof

Soffit boards are sometimes used on the underside of the rafter feet to keep out birds etc. and also to insulate the roof space.

Flush eaves

In this method the rafter feet are cut flush with the brickwork face and the fascia is fixed directly to them.

Open eaves

Cheaper construction methods will sometimes leave the rafter ends completely devoid of either fascia or soffit. The gutter brackets in these cases are fixed direct to the rafter feet.

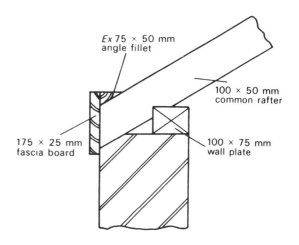

Flush eaves finish

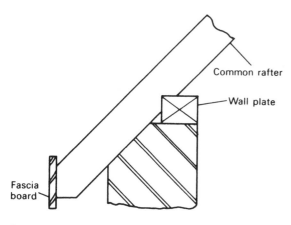

Open eaves finish

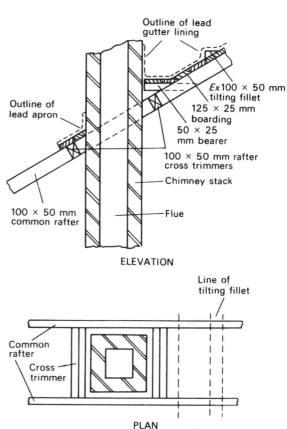

ELEVATION

PLAN

Detail of rafter trimming and formation of chimney gutter

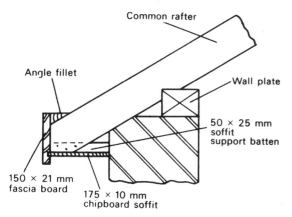

Closed eaves finish

Closed eaves

This is the most common treatment, providing fascia and soffit board. The soffit can be made wider by allowing the eaves to continue in length until the required distance from the wall face is achieved.

Formation of chimney gutters

Where a chimney stack emerges from the slope of a pitched roof it is necessary to form a gutter behind the stack at the highest point and to ensure that the roof covering materials form a water-tight joint on all its sides.

Firstly, the rafters must be trimmed around the brickwork. This is done by cutting and fixing a *trimming piece* behind and in front of the stack. The trimming piece is usually of the same material as the rafters and is fixed at right angles to the slope of the roof. A gap of 50 mm should be left between the faces of brickwork and the trimmers and rafters.

The chimney gutter should be about 100 mm wide. It is formed behind the stack by nailing 25 × 50 mm battens on to the rafters, which provide a fixing for the boarding. A *tilting fillet* is fixed at least 150 mm from the brick face to tilt the tiles into the gutter, which would be zinc or lead lined. Zinc or lead flashing would be inserted to the other faces of the stack.

Trussed rafters

So far we have looked only at the traditional forms of roof construction. It is now necessary to introduce a relatively modern method which is very widely practised in house building: the use of *trussed rafters*. These are made entirely in the factory, transported to site and lifted into position by crane or by manual methods.

Timber sizes are considerably reduced from the traditional roof members. In some spans the thickness of timber can be as little as 35 mm and for larger spans, 47 mm.

The great majority of trussed rafters used on housing developments are of low pitch, but steeper pitches are available if required. Timbers used are all stress graded to avoid the use of defective timber and, unlike traditional roofs, it is entirely wrot (planed) material. There are no conventional joints, the whole structure is butt jointed.

The trusses are assembled on a jig bench and are fastened by *gang nail punched metal plates* which are embedded on both sides of all joints. These are started by hand and pressed home by a hydraulic press. Gang nail plates are galvanised mild steel plates with rows of teeth pressed out over the whole surface. These fasteners are 1 mm thick and vary in size to suit requirements.

On-site storage is important if damage to trusses is to be avoided. Vertical storage, as shown, is the ideal method, but if horizontal storage is necessary sufficient bearers should be used beneath the truss joints to avoid fracture. It is recommended that wall plates should be fixed for this type of roof and wall plate straps are available for this purpose.

With plates secured firmly the trussed rafters can be skew nailed to the plate or fixed with a patent truss clip.

Diagonal bracing must be fixed to the underside of rafters adjacent to gable ends. The angle of diagonals should be 45° in plan. This bracing

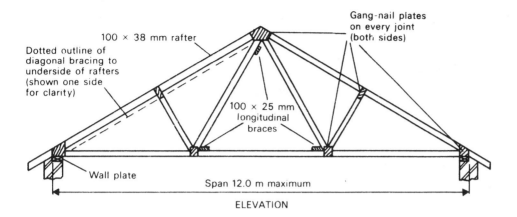

100 × 38 mm rafter

Dotted outline of
diagonal bracing to
underside of rafters
(shown one side
for clarity)

Gang-nail plates
on every joint
(both sides)

100 × 25 mm
longitudinal
braces

Wall plate

Span 12.0 m maximum

ELEVATION

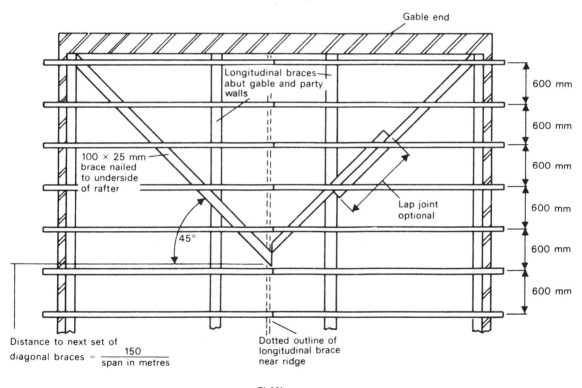

Gable end

Longitudinal braces
abut gable and party
walls

600 mm

600 mm

100 × 25 mm
brace nailed
to underside
of rafter

600 mm

Lap joint
optional

600 mm

45°

600 mm

600 mm

Distance to next set of
diagonal braces = $\dfrac{150}{\text{span in metres}}$

Dotted outline of
longitudinal brace
near ridge

PLAN

Trussed rafters

should be repeated at intervals, the formula for which is 150 divided by the span in metres. In certain cases it is recommended that longitudinal bracing should be fixed through the building. Bracing should be 25 × 100 mm and fixed by galvanised wire nails 65 mm long. Trussed rafters are spaced at 600 mm centre to centre.

The assembly of trussed rafters

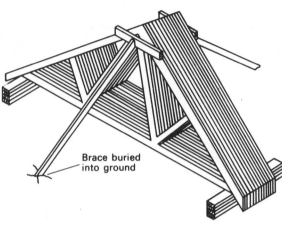

Vertical storage of trusses on site prior to use

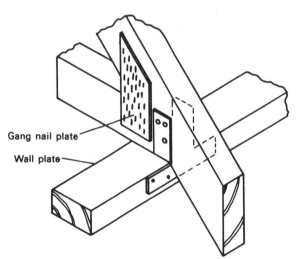

Truss clip (alternative to nailing) can be used on either side of wall plate

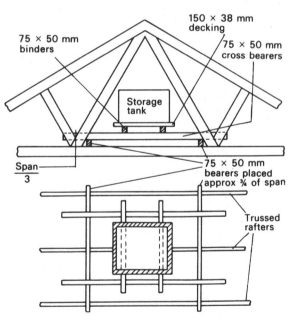

Storage tank platform to distribute weight equally

Storage tank platforms

When provision has to be made for water storage tanks it is necessary to provide a platform made up of 50 × 75 mm timber bearers spread over three trusses. Bearings should be as near to one third of the span from each side as possible. This will distribute the loading across the trusses.

Ceiling hatches

Ceiling hatch openings should be accommodated as far as possible between the spacings of the trusses. This will keep the trimming required down to a minimum. It will be noted that a short ridge and purlins are re-

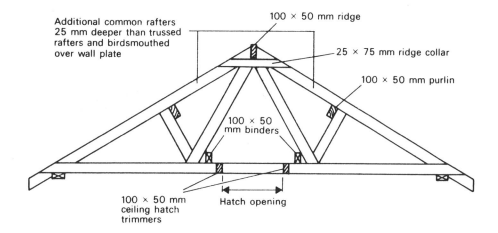

Additional common rafters 25 mm deeper than trussed rafters and birdsmouthed over wall plate

100 × 50 mm ridge

25 × 75 mm ridge collar

100 × 50 mm purlin

100 × 50 mm binders

100 × 50 mm ceiling hatch trimmers

Hatch opening

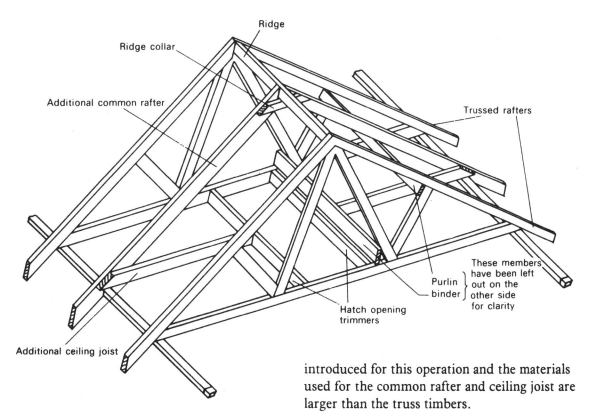

Ridge

Ridge collar

Additional common rafter

Trussed rafters

These members have been left out on the other side for clarity

Purlin
binder

Hatch opening trimmers

Additional ceiling joist

Details of forming ceiling hatch in trussed rafters

introduced for this operation and the materials used for the common rafter and ceiling joist are larger than the truss timbers.

Eaves details vary from those of the traditional roof. It will be noted that rafters are not birds-mouthed over the plate except, as described, when trimming for hatches etc. Temporary bracing is used during erection, but is removed when diagonal and longitudinal braces are secured in position.

Roof trusses

This is an entirely different type of domestic roof truss which, for a variety of reasons, is not used as frequently as the trussed rafter. It is bolted together using timber connectors and retains the use of ridge boards and purlins although these are of much smaller size than in traditional roofs. The trusses are spaced at 1.8 m, centre to centre, with common rafter infilling.

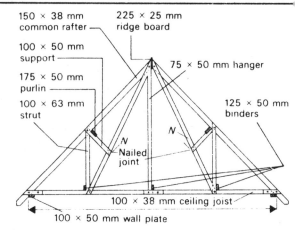

150 × 38 mm common rafter —
225 × 25 mm ridge board
100 × 50 mm support —
75 × 50 mm hanger
175 × 50 mm purlin —
100 × 63 mm strut —
125 × 50 mm binders
N
N
Nailed joint
100 × 38 mm ceiling joist
100 × 50 mm wall plate

Span up to 9.3 m for this type of truss spaced 1.8 m c-c with single rafter infilling. All joints formed with bolts and timber conectors except those marked '*N*'

Roof truss

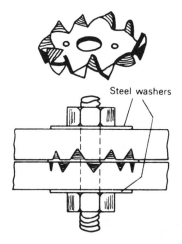

Steel washers

'Toothed ring' wood-to-wood connector

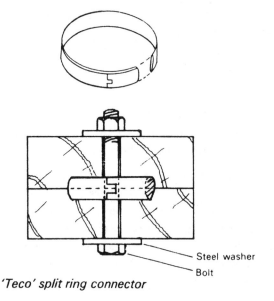

Steel washer
Bolt

'Teco' split ring connector

The two most common and efficient timber connectors used are illustrated. 'Toothed ring' connectors provide a strong connection, comprising a circular steel plate with toothed edges bolted between the timbers. 'Teco' split ring connectors are considered the most efficient of all connectors. The split ring is a circular band of steel split at one point with a 'tongue and groove' type joint. Half the depth of the connector is recessed into each timber and bolts with washers are used to apply the pressure.

Rafters are all birds-mouthed over the wall plate. The trusses are suitable for spans up to 9.3 m and the timber used is all stress graded. It will be noted that some members have shallow notching, this gives support where it is needed to the purlin, and at the upper and lower parts of the verticle struts.

In the construction of this type of truss it is essential that the bolts are placed absolutely

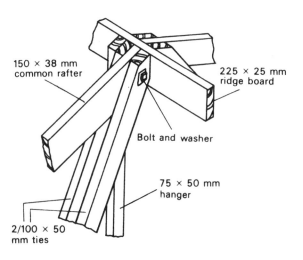

150 × 38 mm common rafter

225 × 25 mm ridge board

Bolt and washer

75 × 50 mm hanger

2/100 × 50 mm ties

Truss detail

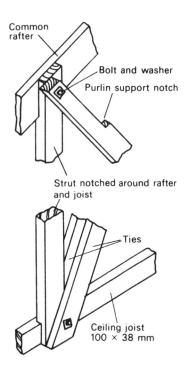

Common rafter

Bolt and washer

Purlin support notch

Strut notched around rafter and joist

Ties

Ceiling joist 100 × 38 mm

Truss detail

central to the positions they occupy. The overall strength of the truss depends on all components playing their maximum part. Therefore, any neglect could result in a weakened structure.

Finding the true lengths and bevels of a pitched roof

Two methods are most commonly used to determine all the true lengths and bevels needed to 'pitch' a roof. It should be remembered that a roof ought to be set-out and cut entirely on the ground before going up on to the scaffold to fix. The two methods are:

1 *Geometry*
   Simple geometrical principles are applied to the size and pitch of the roof, resulting in full accuracy.

2 *Steel square*
   The steel square is designed for any work based on the right angled triangle. The square is favoured by many for setting-out the roof members.

However, before commencing to set-out the roof by any method, the following terms must be learned and thoroughly understood because they are fundamental to the theory of roofing, not least in the setting-out.

Span
The horizontal distance between the outside face of each wall plate.

Rise
This is the vertical height of the roof measured from the upper face of the wall plates to the point of intersection of the pitch lines.

Pitch line
Parallel with the angle of pitch, this line runs

from the outside upper corner of the wall plate to where it intersects with the pitch line on the other slope.

Pitch

The angle, in degrees, created by the roof slope in relation to the horizontal line across the wall plates.

Run

Half the span, (see also rafter run).

Co-run

The distance from the corner of a building to the centre line of the first common rafter in a hipped end.

Note: in a roof which is regular in plan, the run and co-run will measure the same length.

Rafter run

The distance, measured horizontally, that a common rafter 'runs' before reaching its full height.

Rafter length

This is the distance, measured on the pitch line, from the angle of the birds-mouth to the intersection of pitch lines *less* half the thickness of ridge board.

Note: The *eaves allowance* i.e. that which is left on for overhang, must be added to the rafter length.

Hip run

The distance, measured horizontally, that the hip rafter 'runs' before reaching full height.

Hip length

As for the common rafter this is measured on the pitch line and the eaves allowance must be added.

Plumb cut

The angle of cut required at the top of the common and hip rafter. It is also the vertical portion of the birds-mouth.

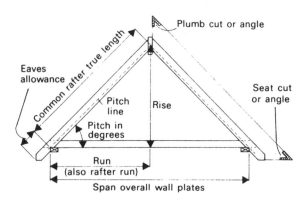

Section through roof

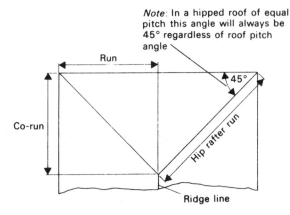

Hipped roof in plan

Seat cut

This is the horizontal cut required to complete the birds-mouth and also to receive the soffit board at the eaves.

Note: A useful check on accuracy can be carried out by adding together the number of degrees for the plumb and seat cut. This should total 90°. Where this is not so an inaccuracy is indicated and remeasurement should therefore be carried out.

The geometrical method

In this method a large scale line diagram must be drawn of the section and part plan (showing at least the complete hipped end). The larger the scale the more accurate will be the lengths and bevels. (Remember that the hip rafters in plan create an angle of 45° and not necessarily the same as the roof slope in section.)

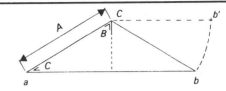

Common rafter section

A = Common rafter true length
B = Common rafter plumb cut
C = Common rafter seat cut
D = Hip rafter true length
E = Hip rafter plumb cut
F = Hip rafter seat cut
G = Hip rafter edge bevel
H = Hip rafter backing bevel
J = Jack rafter edge bevel
K = Jack rafter diminish

Common rafter details

Looking first at the section the true *common rafter length* is shown, *A*, as well as the plumb cut *B* and seat cut *C*. This gives all the necessary information required for a pitched roof with gable ends. For a hipped end roof we must continue by drawing a continuation line through line *b—d* to *e* on the plan.

Hip rafter details

From point *d* set-out the rise *f* on line *d—e*. The triangle *d—f—a* now produces the *hip rafter true length D*, the *hip plumb cut E* and the *hip seat cut F*.

Referring to the diagram showing the intersection of the ridge, two common rafters, two hip rafters and the end common rafter in plan, it will be seen that the hip rafter has a double edge cut to enable it to sit snugly into the angle. To obtain this cut a line is drawn from point *a* at 90° to *a—d* until it touches the extended ridge line *n*.

The hip run *a—d* is extended and the true hip length *a—f* is placed upon it *f'*. A line joining *n* with *f'* gives the angle *G* which is the *hip edge cut*. The hip backing bevel is obtained by drawing a line from point *X* at 90° to the hip *a—f* to intersect with *d*. Then, with *d* as centre and *X* as radius, draw an arc to cut the hip run *a—d* giving point *r*. Point *r* is joined to *b* and *e* revealing angle *H* which is the *hip backing bevel*.

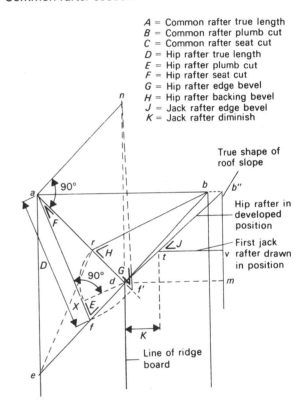

True shape of roof slope

Hip rafter in developed position

First jack rafter drawn in position

Line of ridge board

Hip rafter detail

Jack rafter details

In order to determine jack rafter *bevels* and

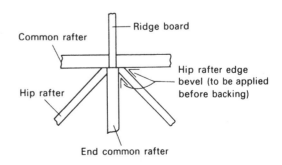

Ridge board
Common rafter
Hip rafter edge bevel (to be applied before backing)
Hip rafter
End common rafter

Intersection of hips with common rafters and end common rafter diminish it is necessary to develop the side of the roof. Making point *c* the centre and *c—b* the radius turn the roof slope into a horizontal position *b'* and project this point downwards to cut the line *a—b* produced at *b"*, and from *b"* draw the vertical line shown. Point *d* is joined in a straight line to point *b"* and this represents the hip rafter in its developed position. When the jack rafter *t—v* is drawn in, the angle between it and the hip is the *jack rafter edge bevel*. The *jack rafter plumb* cut is always the same as the common rafter plumb cut. On the developed side of roof the first jack rafter *t—v* should be drawn in. The distance from the ridge line to where the jack rafter meets the hip rafter is the *jack rafter diminish, K,* and each jack rafter will diminish by this same amount successively.

Purlin details

The edge and side bevels are relatively easy to develop, but the lip bevel (this is where the purlin under-runs the bottom edge of the hip rafter) is a little more difficult and many people prefer to leave this and cut it *in-situ* (on the site). However, all bevels can be obtained accurately if the following rules are followed.

First, draw an over-size (not to scale) purlin in position on the roof section and swing the edge and the side faces into a horizontal plane. Number the original positions 1 and 2 and the developed position 1' and 2'.

Next, project the original points 1 – 0 – 2 vertically onto the plan until the projection lines intersect the hip rafter at 1" – 0 – 2". Points 1' and 2' are now similarly projected until the projection lines intersect with horizontal lines through 1" and 2", respectively. These intersection points are joined to 0, giving as shown: the purlin edge bevel and the purlin side bevel. These intersections are joined back to 0 revealing the *purlin edge bevel*, 3, and the *purlin side bevel*, 4.

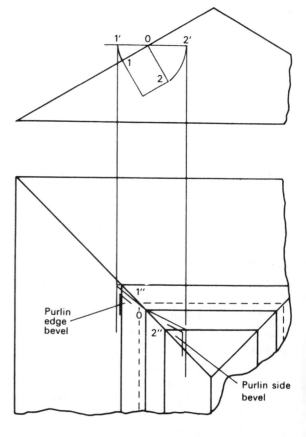

Method of obtaining purlin edge and side bevels

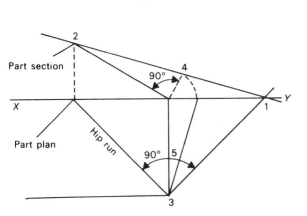

Method of obtaining purlin lip cut

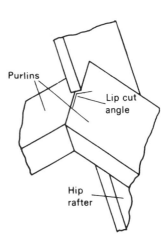

Method of cutting purlins under the hip rafter

To avoid confusion the lip bevel is developed on a separate diagram. A line *XY* is drawn with the half roof section above and the half hipped end below. Draw a line from the corner of the plan 3 at 90° to the hip *r* to give point 1 on the *XY* line. Point 1 is joined by a straight line to point 2 in the roof section. From point 0 draw a line 90° to the roof slope to cut line 2–1 at point 4. An arc is drawn with 0–4 as radius and 0 as centre to touch the *XY* line. A straight line from this point to point 3 will reveal the *purlin lip bevel*, 5.

The steel square method

The use of the steel square in roofing is based on the fact that when any two sides of a right-angled triangle are known, then the length and angles of the third can be calculated. The square has two arms at 90° to each other, the longer and broader arm is called the *body* or *blade* while the shorter narrower arm is called the *tongue*.

Common rafter details
The run and the rise form the right angle of the triangle in which the common rafter is situated, so by setting the rise (to scale) on the tongue and the run (to scale) on the blade the *plumb cut*, 1; *seat cut*, 2; and *true length of common rafter*, 3; are found.

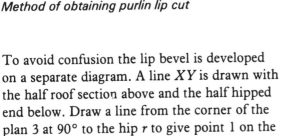

The steel square

Hip rafter details
The hip run is determined by setting the run on the blade and co-run on the tongue, the

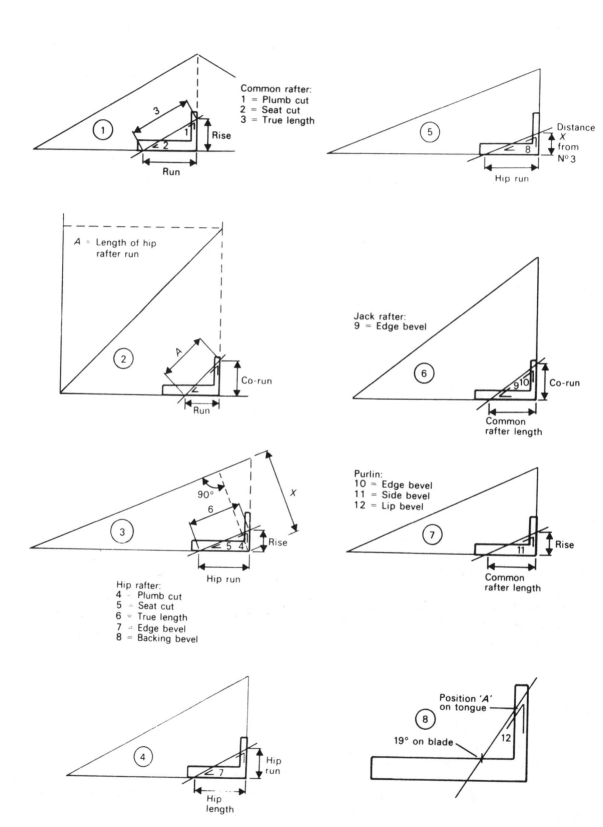

Common rafter:
1 = Plumb cut
2 = Seat cut
3 = True length

Rise

Run

A = Length of hip rafter run

Co-run

Run

90°

X

Rise

Hip run

Hip rafter:
4 = Plumb cut
5 = Seat cut
6 = True length
7 = Edge bevel
8 = Backing bevel

Hip run

Hip length

Distance *X* from N° 3

Hip run

Jack rafter:
9 = Edge bevel

Co-run

Common rafter length

Purlin:
10 = Edge bevel
11 = Side bevel
12 = Lip bevel

Rise

Common rafter length

Position 'A' on tongue

19° on blade

hypotenuse then becomes the *hip run*. With the hip run on the blade and the rise on the tongue the *hip plumb cut*, 4; *hip seat cut*, 5; and *hip rafter true length*, 6, are given. The *hip edge cut*, is found with the hip length on the blade and hip run on the tongue of the square.

The distance X in drawing 3 is now set on the tongue with the hip run on the blade to give the *hip backing bevel*, 8.

Jack rafter details

The plumb and seat cut for jack rafters are the same as for common rafters. To find the *jack rafter edge cut*, 9, the common rafter length is set on the blade with the co-run on the tongue.

Purlin details

With the common rafter length on the blade and the co-run on the tongue the *purlin edge cut*, 10, can be seen (drawing 6). The *purlin side cut*, 11, can be determined with the common rafter length on the blade and the rise on the tongue. The lip bevel is obtained by referring to the tables marked 'Purlin lip cut' on the tongue of the metric steel square. Adjacent to the pitch of the roof are found corresponding degrees. Assuming 35° as the roof pitch the figure 19° can also be seen. This number, e.g. 19°, is set on the inner edge of the blade and a line from 19° passing through centre A on the inner edge of the tongue will reveal the *purlin lip cut*, 12.

Flat roofs

As stated earlier in this section, flat roofs are constructed in a similar way to single floors. They are used quite extensively on extensions or additions to existing buildings. Although they are termed 'flat', there must be a slight fall in the surface of the roof towards the guttering which will take the rainwater away. If the fall is anything less than 1 in 60 there is a danger of water laying on the roof and eventually damaging the roof covering. Therefore, a fall of 1 in 60 is the minimum fall requirement. Tapered pieces of wood, called firring pieces, are fixed to the tops of the joists to create the fall. When the joists run in the direction of the fall, all that is required are tapered firring pieces falling from back to front. Where the joists traverse the fall, it is necessary to cut gradually diminishing firring pieces on each joist, with the widest at the back and thinnest towards the front near the gutter.

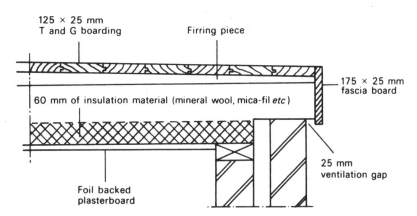

125 × 25 mm
T and G boarding

Firring piece

175 × 25 mm
fascia board

60 mm of insulation material (mineral wool, mica-fil *etc*)

25 mm
ventilation gap

Foil backed
plasterboard

Detail showing flush eaves construction with insulation of roof space

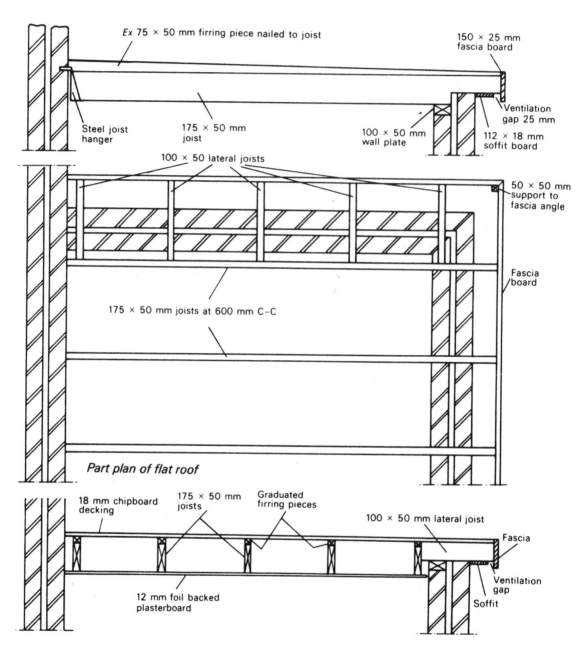

Ex 75 × 50 mm firring piece nailed to joist

150 × 25 mm
fascia board

Steel joist
hanger

175 × 50 mm
joist

Ventilation
gap 25 mm

100 × 50 mm
wall plate

112 × 18 mm
soffit board

100 × 50 lateral joists

50 × 50 mm
support to
fascia angle

175 × 50 mm joists at 600 mm C–C

Fascia
board

Part plan of flat roof

18 mm chipboard
decking

175 × 50 mm
joists

Graduated
firring pieces

100 × 50 mm lateral joist

Fascia

Ventilation
gap

12 mm foil backed
plasterboard

Soffit

Section through roof where joists are parallel to building

The sawn timber joists span from wall to wall, they can be carried by joist hangers (this is common when coming off an existing building) or timber plate bearing. Short lateral joists are used on the sides where necessary to provide fixings for fascia boards. The eaves can be flush or overhang with a soffit. Both methods, however, should be constructed with sufficient ventilation to prevent a moisture build-up in the roof space to comply with the requirements of approved document 'F' of the Building Regulations. Foil-backed plasterboard is fixed and plastered to the underside. Insulation to a thickness of 60 mm is also required in the roof space.

Decking on the top side can be either T and G boards or chipboard sheets. Joist spacings are calculated to suit the decking. 450 to 600 mm spacings are usual, the wider spacing being for thicker decking materials.

Roof coverings

The type of materials to be used for roof covering must be considered in relation to the pitch of the roof. Porous roof coverings should only be used on roofs of fairly steep pitch so that there is no danger of water collecting and causing a breakdown in materials.

Pitched roofs

By far the most common forms of roof covering for pitched roofs are:

1 *Tiles*
For plain tiles the roof should have a pitch of 40° or more.

2 *Slates*
Suitable for roofs of any pitch, because slate is a natural material and totally trustworthy. Not only are various sizes of slate in use, but some are head nailed, e.g. nailed at the top edge, and others are centre nailed.

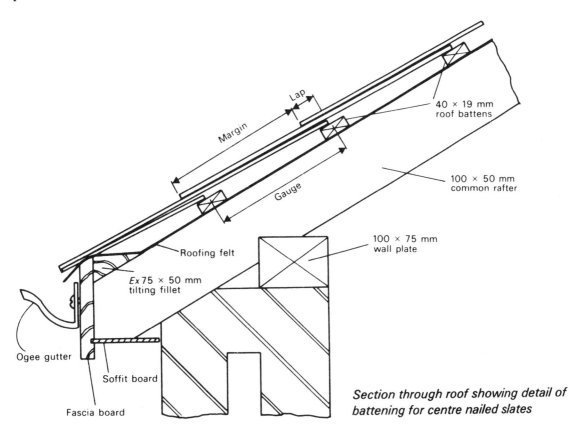

Lap

Margin

40 × 19 mm roof battens

Gauge

100 × 50 mm common rafter

100 × 75 mm wall plate

Roofing felt

Ex 75 × 50 mm tilting fillet

Ogee gutter

Soffit board

Fascia board

Section through roof showing detail of battening for centre nailed slates

Both of these methods require battens to be fixed to the roof first. Battens are usually sawn softwood about 40 × 19 mm and nailed to the rafters at such distances apart as will suit the slates or tiles to be used. This distance is called the *gauge*. The amount that the bottom edge of a slate or tile extends beyond the head of the slate or tile next, but one below it, is called the *lap*. The lap is important when working out the gauge for the slates or tiles to be used. This can be done using the formula (for tiles) of length of tile minus lap required, divided by two. For example a tile 265 mm long is reduced by the amount of lap required, say 65, leaving 200. This figure is then divided by two, which leaves 100. Therefore, 100 becomes the gauge in millimetres for this size of tile.

The *margin* is the exposed area of each tile or slate on the roof. This distance will always be the same as the gauge.

The formula for calculating the gauge for slates is:

1 *Centre nailed slates*

$$= \frac{\text{length of slate} - \text{lap}}{2}$$

$$= \text{gauge in mm}$$

2 *Head nailed slates*

$$= \frac{\text{length of slate} - (\text{lap} + 25)}{2}$$

$$= \text{gauge in mm}$$

The addition of 25 to the lap in head nailed slates is because the actual fixing holes are 25 mm down from the slates' top edge. Battens are fixed on top of a layer of felt for insulation purposes and battening is commenced at the eaves so that any necessary cutting is done at the top and covered by the ridge tile.

Flat roof insulation

As we have seen, the construction of flat roofs is reasonably straightforward but efficient insulation is important for the purpose of energy conservation. To achieve this there are two forms of construction that can be used. These are

1   COLD DECK construction
2   WARM DECK construction.

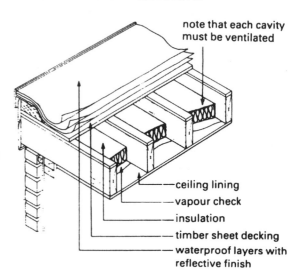

Cold deck construction

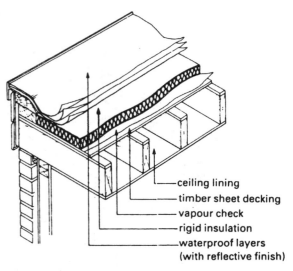

Warm deck construction

In cold deck roofs the insulation is placed immediately above the ceiling board with a minimum of 50 mm air space between it and the decking material. Ventilation must be provided at each end equal to 0.4 per cent of the total roof area. A vapour barrier must also be positioned between the ceiling board and the joists. This method of construction allows for easy maintenance and is, therefore, more widely used.

Warm deck roofs have the insulation placed on top of the joists, making the ventilation requirements unnecessary. However, it is essential to provide a rigid weatherproof membrane on top of the insulation to prevent crushing. The vapour barrier is also required and is situated between the decking and the insulation material. Warm deck roofs are also known as sandwich roofs.

Questions

1   Draw the section through a 40° pitched roof suitable for a 7.5 m span, naming all members of the structure.

2   Using sketches show two methods of fixing sprocket pieces.

3   Using sketches show clearly the formation of:
(a) flush eaves,
(b) closed eaves.

4   Describe the method of manufacture of trussed rafters.

5   State the formula used to determine diagonal bracing positions.

6   Draw a section through a 45° pitched roof to illustrate the following terms:
(a) span,
(b) rise,
(c) pitch line.

7   What is the minimum fall recommended for a flat roof? Give reasons.

8   Describe two methods of insulating flat roofs.

10 Timber stud partitions

This is a very common form of partitioning because it is simple to construct and yet very effective.

Partitions are used to divide large rooms into smaller rooms, or to form lobbies near entrance doors. They are also found purely as screen walls in many cases. Partitions are erected with or without door openings, with or without glazed openings, and can be straight, splayed or right angled in plan. These types are all non-load-bearing – they have no structural function. The load is carried by the main walls of the building and any internal wall which is constructed in load bearing materials such as brick.

Construction

Timbers used are normally 50 mm thick sawn softwood with widths varying from 75 mm to 100 mm according to requirements. The main components of the partition are the *head plate*, *floor plate* and *wall pieces*. All these members must be accurately and firmly fixed if the whole structure is to remain rigid. The floor plate would be screwed down first, followed by the fixing of the head (or ceiling) plate. Where the floor and ceiling joists are running at right angles to the line of partition there is little problem in getting ample fixing. However, noggings between the joists become necessary when the joists and the partition run parallel. For adequate strength these noggings should not exceed 600 mm spacing.

Wall pieces must be plumbed-in and fixed firmly to the wall if necessary with rawlplugs and screws. These are infilled with vertical timbers called *studs* and the short pieces of timber cut between the studs are called *noggings*. The noggings serve two purposes. They provide intermediate fixings for the cladding material and assist in making the partition generally more rigid. The studs can be forked over a batten screwed to the head and floor plates, as shown, to add firmness to the partition. Most timber stud partitions are clad with sheet material of 1220 mm standard width.

It is very important to consider the size of the cladding material before erecting the partition because correct spacing of the studs and noggings can keep the cutting and waste of material to a minimum. If sheet materials 2440 × 1220 mm were being used, for instance, the first stud should be spaced 1220 mm from the wall face to the centre of the stud and succeeding studs should be spaced at 407 mm centre to centre.

Additionally, a line of noggings would be placed at 2440 mm centre above the floor to give the boards edge a fixing and other rows of noggings would be spaced equally between the floor plate and this row. Noggings are fixed by skew nailing in a continuous line or staggered, the latter being easier to nail.

Where one stud partition is abutted by another, either at a corner or somewhere in its length, it is sometimes necessary to include additional studding and common methods of doing this are shown. However, when all the work is carried out in one operation it is possible to eliminate one stud by cladding the first partition before erecting the partition which runs into it. This is shown in the illustration.

Partitions which contain door openings, lights or other forms of opening require trimming in order to retain the strength and rigidity of the partition.

All such openings will have a lining frame, the thickness of which must be taken into account when calculating the actual opening in the stud work. It is recommended that in addition to the lining thickness, a further 25 mm gap should be allowed for between the back face of the lining and the studs. This will allow folding wedges and packing to be inserted when fixing the

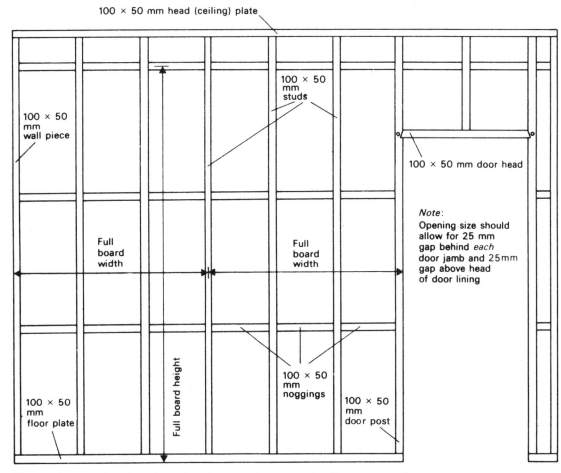

(Note positioning of studs and noggings to economise on board cutting and waste)

Elevation of simple stud partition with door opening

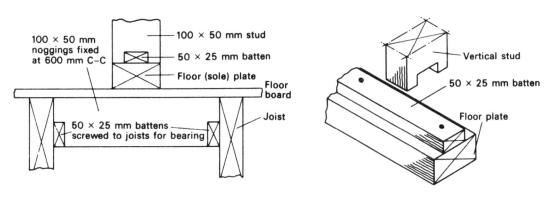

Method of forming studs over batten screwed to floor plate

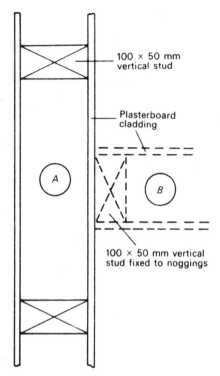

Plasterboard cladding

100 × 50 mm vertical stud

100 × 50 mm vertical stud fixed to noggings

Note: Dotted partition 'B' must be fixed *after* partition 'A' has been fixed and clad

Plan of stud arrangement which uses less timber

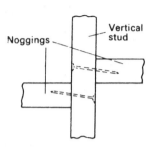

Method of staggering noggings between studs for easy fixing

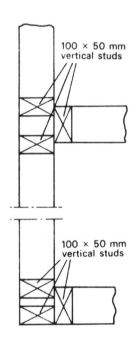

Arrangement of studs in plan at angles and where partitions intersect

linings to ensure that they are absolutely straight and square. The width of the lining must also be sufficient to allow for the thickness of the cladding material.

Architraves, usually of simple design and light section, are applied to the face to mask the joints left between the cladding and linings. Glazing to any light openings is usually direct to the lined opening and beaded. Where desirable it can be double glazed for insulation purposes.

The stud work at each side and across the head of a door opening must receive special attention. To achieve maximum strength at the head the *head piece* should not only be tenoned into the vertical studs, but should also be fitted

with splayed shoulders. Dowels running through the joint should ensure maximum hold.

The foot of each side stud should be fitted to the floor plate on each side with an open mortise and tenon joint and dowelled as at the head joint. Much stress is borne by the joints at

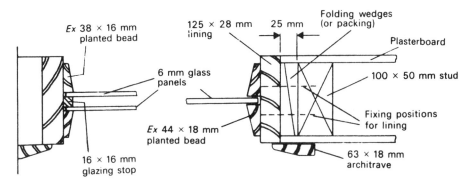

Details of single and double glazed opening in stud partition

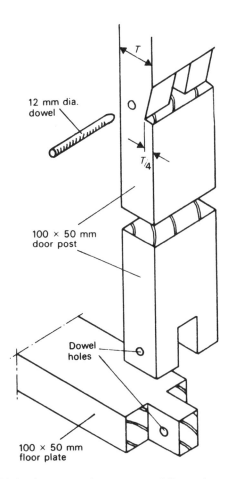

Detail of joint between door post and floor plate

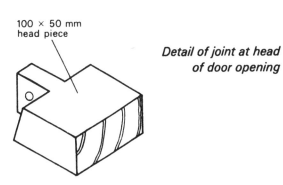

Detail of joint at head of door opening

this position and failure to make a sound job will almost certainly result in the door lining becoming loose in the course of time.

Insulation

It is sometimes necessary to construct timber stud partitions in such a way as to minimise the transmission of sound and heat. This is called *sound and thermal insulation*.

There are many materials available for use in this type of insulation work. Whatever method is used, however, the stud work must be erected with at least one thickness of insulation material, e.g. felt or insulation board, between the floor and floor plate, between the head plate and ceiling and between each wall piece and the wall. Without these any other precaution taken would be seriously undermined.

The staggered stud method is very effective although it is more costly because of the use of additional materials. Its effectiveness arises from the fact that a definite break exists in the two sides of the partition.

Other methods of construction are shown using insulation boards and insulation quilt. Although these methods are fairly common they are by no means exhaustive. Double glazing can be used where lights occur in a partition. The wider the cavity between the panes of glass the more effective an insulator it becomes.

Hollow core flush doors are poor insulators. Therefore, doors of more substantial construction must be used, including solid core flush doors.

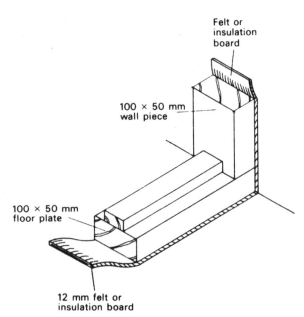

Treatment around partition to reduce passage of sound

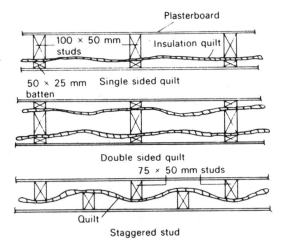

Various treatments of stud partition for insulation

Questions

1 What governs the spacing of studs?

2 State the purpose of noggings.

3 Name the type of nail used in partition work.

4 What advantage is gained by staggering noggings?

5 Why are stud openings made larger than the actual lining size?

6 How is the passage of sound and heat reduced in stud partitions?

7 Describe the sequence of operations used in the erection of a stud partition.

8 Describe a method of giving added strength to the top and bottom of the door posts.

11 Stairs

Staircases are provided in buildings to give access to floors which are on different levels. The staircase in a private dwelling which is used mainly for those living in that dwelling is known as a *private stairway*. Any staircase, whether internal or external, which is for the use of two or more dwellings is called a *common stairway*. It is while ascending or descending a staircase that accidents are likely to occur and it is for this reason that the design and construction regulations relating to stairs are so stringent.

Terms associated with stairways

In this section we shall concern ourselves with the design, construction and fixing of private stairways only, but before doing so the following terms used in staircase work should be clearly defined and understood:

Wall string
The wide outer board of the stairs which is fixed to the wall adjacent to it.

Outer string
When a staircase is open on one side and against a wall on the other the outer string is opposite to the wall string.

Tread
The horizontal board on which the foot is placed when using the stairs.

Riser
This is the vertical board between any two treads.

Step
Every assembled tread and riser in a traditional timber staircase is called a step.

Going
The horizontal distance measured on plan from the edge of nosing on one tread to the edge of nosing on the next tread.

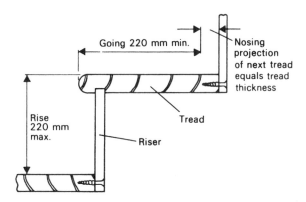

Section through a step showing definitions and regulations

Rise
The vertical distance measured from the top surface of one tread to the top surface of the next tread.

Newel

Large posts, usually 100 × 100 mm, into which the outer string is tenoned. They are capped at the top either with a separate newel cap or by being chamfered or rounded etc. in the solid.

Handrail

A rail fixed in a position to give assistance to those using the stairs. They are sometimes tenoned into the newels or fixed to the wall.

Baluster

The name given to the rails which are fixed under the handrail on an outer string and on the top edge of the outer string.

Balustrade

This is the area bounded by the inner edge of the vertical newel posts and the top edge of the outer string up to the underside edge of the handrail.

Flight

A straight set of steps of any number.

Fliers

These are the parallel steps in a straight flight of stairs.

Pitch line

An imaginary line which runs at the pitch of the stairs touching the nosing of each step.

Building regulations

In addition to these terms the requirements of the Building Regulations (Approved Document K) must also be studied and complied with. The main regulations for private stairways are as follows:

1 The maximum *pitch* must be 42°.
2 The *going* of any step must not be less than 220 mm.
3 The *rise* of any step must not be more than 220 mm.

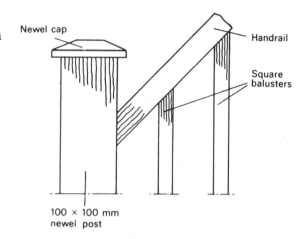

Newel and part of balustrade

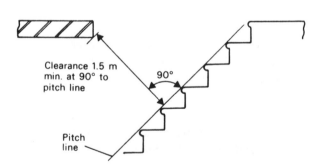

Clearance regulations

4 For any *flier* the sum of the going plus twice the rise must not exceed 700 mm, nor be less than 550 mm.

Bearing the above points in mind, when calculating the rise and going of a staircase the following formula can be used:

$$2R + G = 550 - 700 \text{ mm}$$

where R = rise and G = going.

5 The *headroom* must not be less than 2 m measured vertically above the pitch line.

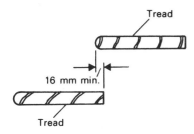

Minimum overlap of treads

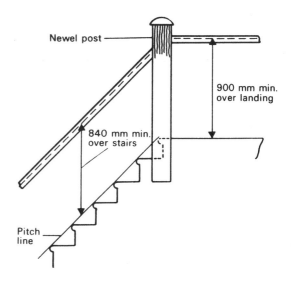

Minimum handrail heights

6 The *clearance* must be a minimum of 1.5 m measured at 90° to the pitch line. This provides a clear way parallel to the stairs throughout their entire length.

7 The nosing of the tread or landing must overlap the back edge of the tread below it by at least 16 mm. This applies particularly to stairs with open risers.

8 Any stairway which is less than 1 m wide must have one handrail and two handrails if it is more than 1 m wide.

9 Handrails shall not be less than 840 mm vertical height on the pitch line and not less than 900 mm above the landing floor.

Design

We shall assume that a dwelling is to be fitted with a straight flight of stairs which will be fixed to the wall on one side, with an open balustrade on the other.

The first thing that must be ascertained is the *total going* and the *total rise* and, if possible, these should be taken on-site. The total going is the horizontal distance between faces of the first and last risers, while the total rise is the vertical distance between the finished floor line (FFL) at the lower level and the FFL above. Provided there are no other restrictions, the total rise is the important measurement because it must be accurately divided into equal rises on each step or flier, and should be marked onto a *storey rod*

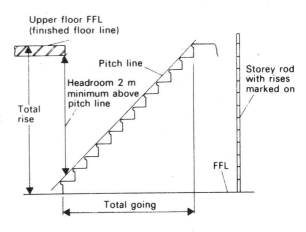

Straight flight with definitions and regulations

made from straight batten of small section. For example, a total rise of 3.215 m, using the regulation formula, would be divided by 220, which is the maximum rise allowed on a step. As this works out at 14.6 it is clear that 15 rises will be necessary and, therefore, 3.215 m is divided by 15 to give a rise of 214 mm.

At this point it should be observed that the number of treads in a flight will always be one less than the number of risers. This means that, with 15 risers, 14 treads will be required. Assuming our total going measured 3.640 m it will be seen that this figure divided by 14 gives an answer of 260 mm. Now, using the whole formula: $2R + G = 550 - 700$, we can see that the rise and going of our imaginary staircase, e.g. 214 mm rise and 260 mm going are within the regulations.

Construction

This information above is the starting point from which the stairs are set-out. The materials must now be drawn together and construction commenced. Work will normally begin on the strings. Two boards of 32 mm thickness and wide enough to accommodate the treads and risers should be set-out accurately. The wall string has on it all treads and risers and should therefore be marked-out first.

A pitch board made from plywood is cut accurately to the size of tread and rise and is used in conjunction with a margin templet. The *margin* is the parallel distance from the top edge of the string to a line running through the angle of each step before the nosing is applied. This distance is optional. A narrow strip of wood cut to the width of the margin and about 300 mm long is then fixed by pins to another strip to form a 'T' or a simple rebated piece. This is used against the string edge so that the pitch board can be placed against it to outline the treads and risers.

Using the pitch board and margin templet the floor line is drawn at one end of the wall string and each subsequent riser and tread is drawn in outline only until riser number 15 is reached.

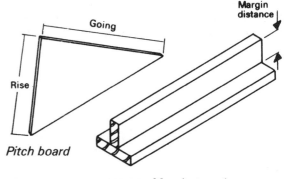

Pitch board

Margin templet

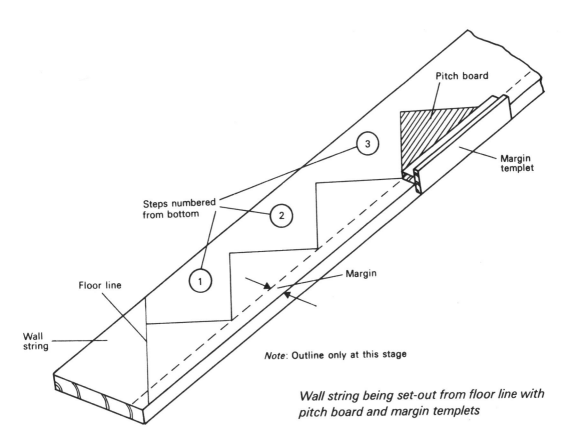

Wall string being set-out from floor line with pitch board and margin templets

Before completing the marking-out a tread templet and a riser templet should be cut from either plywood or hardboard. These must allow for the thickness of the tread and riser as well as the allowance for the wedges which will hold the staircase together. With the accurate templets held against the outline of the treads and risers the marking-out can be completed on the wall string.

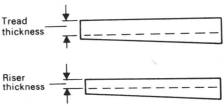

Tread and riser templets cut from ply or hardboard with wedge allowance

The outer string is marked out similarly, but always remember that it must be paired, e.g. the opposite hand, to the wall string.

At this stage the shoulder lines should be marked-out where the tenons occur at the lower and upper end of the outer string. Because of the width of the string at these points a pair of tenons is formed together with haunchings in the manner shown.

When all marking is complete the trenching of the strings is commenced. Trenching is normally 12 mm deep and is carried out either by hand or with a heavy duty portable router. Accuracy in trenching with a flat, equi-distant bottom is essential for good quality. If a hand operation is used it will be found easier if large diameter holes are bored to give a start to the trenching of each tread and riser.

Setting-out completed with tread and riser templets

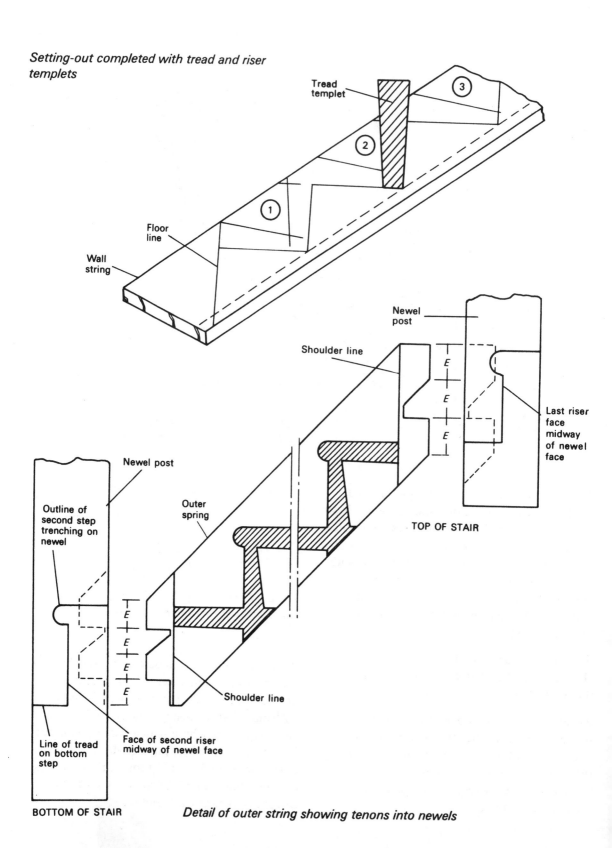

Tread templet

Floor line

Wall string

Newel post

Shoulder line

E
E
E

Last riser face midway of newel face

TOP OF STAIR

Newel post

Outline of second step trenching on newel

Outer spring

E
E
E
E

Shoulder line

Line of tread on bottom step

Face of second riser midway of newel face

BOTTOM OF STAIR

Detail of outer string showing tenons into newels

Step construction

A number of ways exist for the construction of the steps, probably the most common being a 25 mm thick tread with a 12 mm plywood riser housed full thickness into the underside. The *nosing* is the distance from the most forward edge of the tread and the face of the riser. It is usually rounded or shaped and it projects in front of the riser face by the thickness of the riser. Other step formations are shown, but these are not exhaustive.

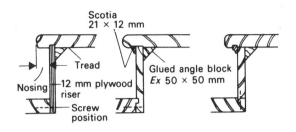

Various forms of step construction

Stairs with no risers are also frequently used nowadays. In this type of construction the tread thickness should increase to a minimum of 32 mm because they are not supported by a riser. When the treads and risers are assembled a jig called a *cradle* is used so that when the glued angle blocks are fixed the tread and riser are at 90° to each other. Glued angle blocks should not exceed 400 mm spacings.

The assembled treads and risers must be stacked carefuly until required. For maximum accuracy each step should be numbered and fitted individually into the strings.

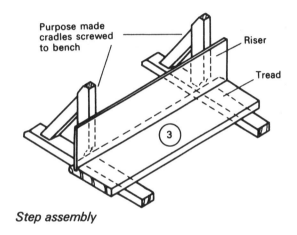

Step assembly

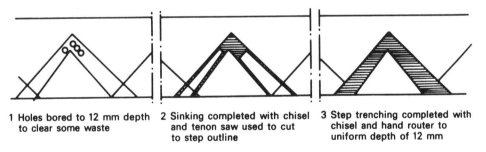

1 Holes bored to 12 mm depth to clear some waste

2 Sinking completed with chisel and tenon saw used to cut to step outline

3 Step trenching completed with chisel and hand router to uniform depth of 12 mm

Sequence of trenching strings when using hand tools

Assembly

With all work completed to the strings and fliers and with wedges cut to shape, the assembly can be carried out. This is done on a sturdy bench which must be flat and level. It is essential for the strings to be absolutely straight and parallel: the bench may have timber or metal attachments to enable assembly to be carried out with ease and accuracy.

Wedging must be done methodically. It will be found easiest to commence at riser number one and wedge both sides. Next, the tread is wedged on both sides and then the second riser, and so on to completion. Remember that it is the *wedges only* that are glued in assembly.

Finally, the riser and the back of the tread are screwed to prevent movement when the step is trodden on.

The handrail and balustrade are now fitted to the newel posts, which have already been fitted to the outer string.

Handrails

The handrail itself can be of simple or moulded section, the important thing being that it is of a size and shape that can be gripped easily. The ends of the handrail are tenoned into the newel at each end at not less than 840 mm above the pitch line to the centre line of the handrail. It is good practice to sink the shoulders of the handrail into the newel to a depth of 6 mm to ensure that a good joint results.

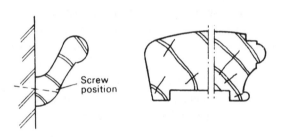

*Wall handrail —
for stairs between walls*   *Plain and moulded
handrail sections*

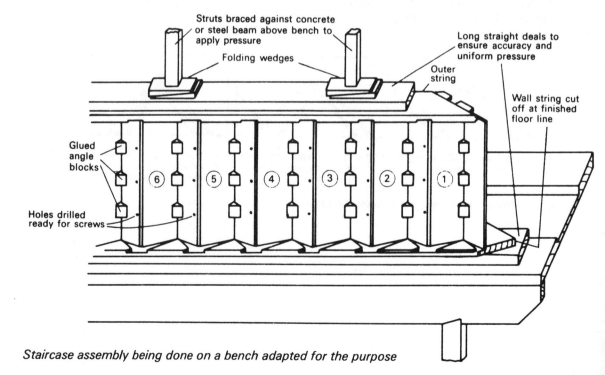

Staircase assembly being done on a bench adapted for the purpose

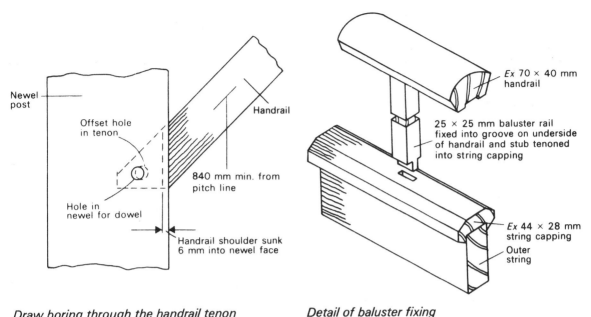

Draw boring through the handrail tenon

Detail of baluster fixing

Draw dowels of 10 mm diameter will hold the joint, but these should not go right through the newel. It will suffice to bore about three quarters of the way through. Draw boring consists of boring the hole in the tenon about 3–4 mm nearer to the shoulder so that when the dowel is driven in the shoulder is pulled up to its maximum. The dowel should be slightly tapered at the front to give it a lead-in when being inserted.

Balusters

Balusters can be square or rectangular in shape and there are others which can be turned on the lathe. When these types of baluster rail are used they should under no circumstances have spaces between them which exceed 90 mm. The balusters will be strong and secure if they are stub tenoned into the string capping at the bottom and run into a groove on the underside of the handrail section at the top.

The balustrading often consists of plywood panelling with muntins grooved to receive the

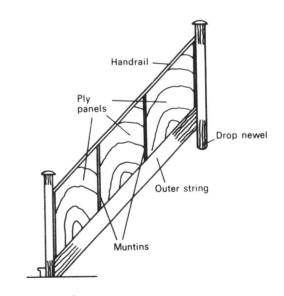

A panelled balustrade

panels. Alternatively, a ranch style combined handrail balustrade is frequently used, but once again the distance between them should not be great enough for children to fall through.

Landings

Where a staircase has to change direction because of confined space, a landing space is sometimes introduced. A staircase which turns once only at right angles has what is called a *quarter space landing*. Those staircases which turn back on themselves are called *dog leg stairs* and have a *half space landing*.

The length of any landing must be at least the width of the staircase itself. Nosings are formed on the leading edge of the landing which will be identical to those used on the fliers. All landings must be supported on a system of joists to ensure rigidity. This is usually, constructed of similar materials to the upper floor. However, the joist against which the stair first makes contact must be a trimming joist (75 mm thick), other bridging joists are used across the landing.

Apron boards of either ply or solid timber are used as a facing to the landing joists.

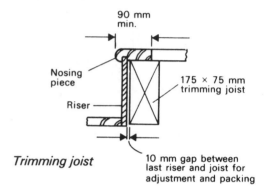

Trimming joist

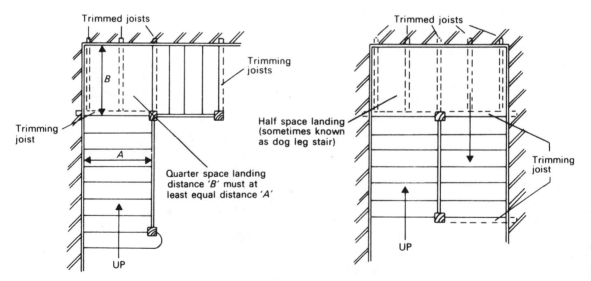

Plans showing types of stair to overcome space restrictions

Site fixing

The fixing of a staircase on-site poses different problems on different sites. Assuming that the straight flight considered earlier in this section is the one to be fixed, it will mean that some fixing through the underside of the wall string will be possible. With thermalite blocks, which are now widely used on internal walls, this fixing could well be done with cut nails, provided that the string is drilled in each fixing position first. As we saw in the section on upper floor construction the stair-well will be properly trimmed and it will be over this trimmer that the upper newel will be notched and screwed. A small gap of 10 mm should be left between the trimmer face and the back face of the last riser. No tread is fitted to the last riser, instead a nosing piece which must not be less than 90 mm (to comply with British Standard 585) is fitted adjacent to the rest of the floor covering on the landing.

Where possible, the bottom newel should be carried through the floor covering and bolted to the joist to give maximum strength to the staircase. In cases where the staircase is 1 m or

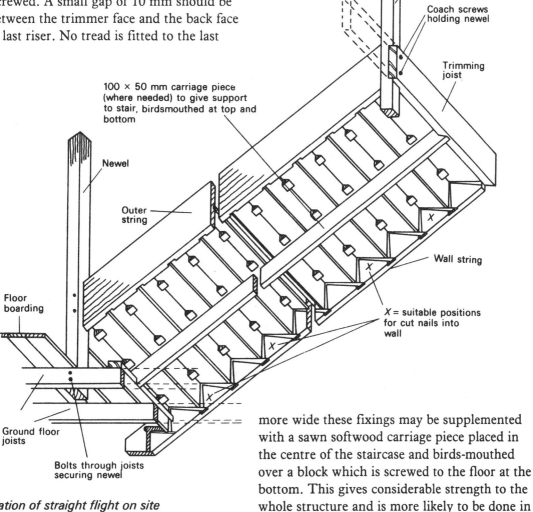

100 × 50 mm carriage piece (where needed) to give support to stair, birdsmouthed at top and bottom

Newel notched over trimming joist

Coach screws holding newel

Trimming joist

Newel

Outer string

Wall string

X = suitable positions for cut nails into wall

Floor boarding

Ground floor joists

Bolts through joists securing newel

Installation of straight flight on site

more wide these fixings may be supplemented with a sawn softwood carriage piece placed in the centre of the staircase and birds-mouthed over a block which is screwed to the floor at the bottom. This gives considerable strength to the whole structure and is more likely to be done in cases where the spandrel is to be panelled or filled in.

The *spandrel* is the triangular area underneath the stair string which is formed by the floor and the underside of the string.

Damage prevention

Having fixed the new staircase on-site, measures should be taken to prevent it becoming damaged. Off-cuts of hardboard, which is very hard wearing, can be pinned to the top of each tread making certain that the front edge near the nosing is well protected.

Hardboard is also used around the newel posts particularly to protect the corners. Strips can be cut to the width of each newel side and held in position with masking tape. However, if the newels are to have a paint finish there is no reason why the protective pieces should not be pinned.

The handrail and balustrade can be adequately protected by corrugated cardboard which, again, can be held in position by masking tape or other types of adhesive tape.

Questions

1 Calculate a suitable going on the flier of a flight of stairs of which the rise is 180 mm.

2 What is the minimum headroom allowed in the Building Regulations?

3 Define the margin on a flight of stairs.

4 Describe the use of the following when setting out stairs:
(a) margin templet,
(b) pitch board.

5 Describe the sequence of operations when assembling a straight flight of stairs.

6 State the minimum heights for handrails:
(a) over the pitch line,
(b) over the landing.

7 Describe the assembly and function of a cradle.

8 Describe the method of fixing a straight flight of stairs to a block wall.

9 When is the use of a carriage piece necessary and what is its function?

10 Describe methods of protecting a newly installed flight of stairs while other work is in progress.

11 Why is a gap left between the trimmer and the last riser in stair construction?

12 What is the best type of fixing for the bottom newel?

13 Describe the method of finishing on the top riser adjacent to the landing floor.

12 Adhesives

Several groups of adhesives (glues) have been developed for the many uses to which they are now put in the Woodworking Industry.

As we saw earlier in the book plywood manufacturers use a variety of adhesives to produce boards of varying quality. There is also extensive use of adhesives in the Laminated Beam and Truss Manufacturing Industry.

Modern adhesives have enabled the use of timber in situations, shapes and sizes that were unheard of many years ago. In many instances, the glued joint is stronger than the timber itself, provided that the job has been done properly.

For this reason it is very important that the woodworker has knowledge of the various groups with their characteristics and uses so that the correct glue can be used for the work in hand. For example, it would be useless to use a glue which has poor resistance to moisture on an external frame or door.

Common terms

Curing time
The period of time from when the glue is used to the time when it has reached maximum strength.

Pot life
This is the length of time that the glue, when mixed, will remain good for use before starting to lose its quality. Adhesives are not safe to use after the expiry of the pot life.

Shelf or storage life
The period during which a glue can be stored before use is called the shelf or storage life. It is unwise to use any adhesive which has been stored for longer than the recommended time.

Adhesives can be divided into five main groups:

1   Animal glue
2   Casein glue
3   Poly-vinyl-acetate (PVA)
4   Synthetic resins
5   Solvent rubber

Animal glue

This is more commonly known as *scotch glue*. The basic ingredients are extracted from the bones and hides of animals. Some types are available in crystal or gelatine form and must be heated in a glue pot before use. However, there is also a variety which can be used cold straight from the tin.

The use of scotch glue has declined in recent years, it having been replaced by modern adhesives which are more convenient to use.

The glue is applied with a brush. Excess glue can be removed with the brush and hot water.

Brushes should be cleaned after use with hot water.

Pressure is required initially to bring the joint up and, depending upon temperature, it will take between 4 and 8 hours before cramps or other forms of pressure can be released. At least 12 hours should elapse before any joint is worked on.

This type of glue has poor resistance to heat or water and should therefore only be used internally. However, it does produce a strong joint with very good tensile strength.

Casein glue

Milk is soured, dried, pressed and ground to a powder to produce this glue. The adhesive forms after the powder is mixed with water and the water evaporates. When mixing, the powder should be left to dissolve in the water for about 15–20 minutes, giving an occasional stir.

Application is by brush, but this glue has a tendency to stain, particularly on hardwoods. It will cure in 2–4 hours. Pressure is required for the initial tack, but 8 hours should be allowed before working on the joint.

The pot life of this glue is 8 hours maximum. Therefore, it should be changed after this time. Casein gives a good strong joint, but because of its low resistance to water and heat it should not be used in fully exposed positions.

Poly-vinyl-acetate (PVA)

This glue is the most widely used at present, not only for woodworking, but as a bonding agent in plaster and sand and cement surfaces.

It is a white plastic emulsion which comes in containers ready for use although it can be thinned with water if required. Thinning with water does, however, reduce its strength. Brushes and excess glue can be cleaned thoroughly with water. It is advisable to apply PVA glue to both surfaces where possible and, because of its low initial tack, joints will require cramping or other forms of pressure for up to 3 hours, according to room temperature.

Pot life presents few problems as this type of glue will retain a good consistency for long periods provided that the container is covered when not in use.

PVA is an easy glue to use because it gives a fairly long assembly time which is useful in work where there are a number of tenons or other joints to assemble. The durability of the glue is very good; and a strong, permanent joint will result. However, it is strictly an internal adhesive because it will break down in moist conditions.

Synthetic resins

It is in this group that the great advances have been made which have led to the introduction of laminated timber buildings, among other things.

The resins used are known as *thermo-setting* because, once set, they are not affected in any way by heat. It is the reaction of certain chemicals which are used to convert the resin from the liquid state to the solid and, for this reason, the manufacturer's instructions regarding the amounts of resin and hardener to be used should be strictly followed. Both resin and hardener must be measured *by weight* and not by volume.

The liquid hardeners are harmful to the skin and dangerous to inhale. Therefore, precautions such as wearing gloves and the use of barrier creams are important.

Those synthetic resins most common to woodworking are:

1   Urea formaldehyde (UF)
2   Phenol formaldehyde (PF)
3   Resorcinol formaldehyde (RF)
4   Melamine formaldehyde (MF)

The resin and hardener are usually mixed according to the manufacturer's instructions and applied by brush. However, in some instances the hardener will be applied to one surface and the resin to the other.

Some of the synthetics have gap filling qualities up to a maximum of 1.25 mm, after which loss of strength occurs.

Synthetic resin adhesives have excellent properties of both durability and strength, as well as having very high resistance to water and heat.

Solvent rubber

These are more commonly called 'contact adhesives' and are used almost exclusively on non porous materials such as plastic laminates. This type is derived from *neoprene rubber* which is combined with a solvent called *acetone*. It is the solvent which permits the even spreading of the adhesive using a scraper with a serrated edge or a brush. Contact adhesives are supplied ready for use and cannot be diluted or thinned.

Having applied the adhesive to both surfaces which are to be bonded, it is then left for 15–20 minutes to become 'touch dry' before bringing the surfaces together. The two dry films 'grab' on contact and pressure can then be applied to increase the adhesion.

In difficult situations when cutting around pipes etc. is necessary, it will be found easier to place strips of narrow hardboard at about 300 mm spacing on one surface. The other surface can then be laid on these strips which can be withdrawn one at a time as pressure is applied. More accurate positioning is possible using this method since no movement can occur once the surfaces have been brought together. Tolerance should be allowed for cleaning off when bonding plastic laminates to timber or chipboard worktops etc. Benzene can be used to clean brushes or other equipment used with this adhesive.

Solvent rubber adhesives do not have great holding qualities and are suitable for light work only. They have little resistance to water and heat. During use a very heavy vapour is emitted and adequate ventilation must be provided to prevent operatives from being overcome by these fumes. There is also the risk of fire in this situation.

British Standard Specification (BSS classification)

Adhesives are classified under a British Standard Specification scheme according to their properties. This system of grading means that materials carrying a grade-mark can be used with confidence in given situations, especially where the durability of the adhesive is important. A summary of the grades used is as follows:

1 WBP

The WBP grade-mark indicates that the adhesive used is *weather and boil proof* as well as being resistant to micro-organisms. Examples are resins of the resorcinol (RF) or phenol (PF) types. Such adhesives are suitable for all exposed conditions.

2 BR

This grade has good *boil resistance*, but is likely to break down after long periods of exposure. Melamine (MF) or urea (UF) are in this group. They are very suitable for external use provided that they are not totally exposed for the whole time.

3 MR

These adhesives are *moisture resistant* with good weather resistance, but are likely to break down in boiling water. Urea (UF) belongs in this category. It is suitable for unexposed external situations and also has wide general use.

4 INT

This indicates *internal* use only and is not for use in damp or moist conditions. The adhesives in this category have very good durability and strength and are suitable for most of the joinery which is for internal use. Animal glue, casein glue, PVA and solvent rubber adhesives all carry this grading.

Questions

1 List five groups of adhesive.

2 Define the following terms:
   (a) curing time,
   (b) pot life,
   (c) shelf life.

3 Why should scotch glue only be used on internal work?

4 List *four* advantages gained by using PVA glue.

5 Why should protective clothing be worn when using some synthetic resin glues?

6 Why are solvent rubber adhesives also known as contact adhesives?

7 List three safety measures to be taken when using solvent rubber adhesives.

8 What does the abbreviation WBP stand for?

9 List four glues which are for internal (INT) use only.

13 Joinery fitments

A great deal of the joinery used today is mass produced. However, there is still a demand for some which has to be purpose-made. These are usually items of joinery where, perhaps, only one fitment is required. It is these 'one off' items which are dealt with in this section.

Particle board, e.g. chipboard, is now used extensively in the manufacture of all forms of joinery and there are instances where it is as good as any other material. It should not, however, be regarded as a suitable replacement for solid timber or blockboard in every case.

Kitchen units

Wall and floor units are constructed using manufactured boards such as blockboard, laminboard or chipboard. In many cases a plastic laminate is used to cover the surface areas. Drawers, doors and shelves are included in most designs and, because of the heavy use to

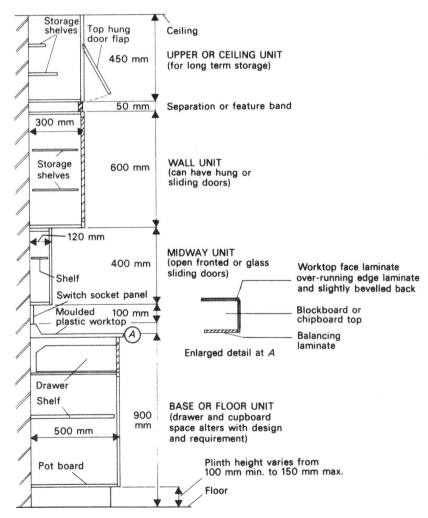

Vertical section showing floor to ceiling units used in modern kitchens

which they are put, all components should be properly constructed so that joints will not break down during use.

Where plastic laminate is applied to any surface there must also be a balancing laminate on the opposite face. This is to prevent distortion taking place which would make the panel become bowed or rounded.

Particular care should be taken to ensure that plastic laminate on worktops over-run the edging strip and have a very small bevel applied which will not only make the corner less dangerous, but will also reduce the possibility of it being lifted by something brushing against it accidentally. A small upstand is also fixed at the back of the top surface against the wall. This serves the dual purpose of covering the joint between the fitment and the wall as well as, making the worktop easier to clean.

Drawer construction

The size of the drawers will vary according to the use to which they will be put. However, drawer depths of between 100 and 150 mm are widely used. Those drawers which are used for cutlery should be lined throughout with felt or baize. Divisions of about 10 mm thickness can also be incorporated to separate knives, forks etc.

Drawer fronts are normally slightly thicker than the sides and backs. This is because it is subject to frequent handling and use. It also enables the use of lapped dovetails between the angle of the front with each side. It will be noted that the drawer back can be housed into the sides provided it is glued and pinned. This method is frequently used. Where dovetails are specified they should be of the plain type and care should be taken to ensure that the dovetails occur on each drawer side with the pins on the drawer back.

The drawer bottom should be grooved into the

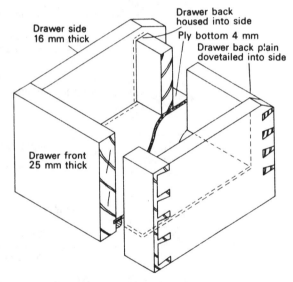

Drawer side 16 mm thick

Drawer back housed into side
Ply bottom 4 mm
Drawer back plain dovetailed into side

Drawer front 25 mm thick

Drawer construction showing alternative treatment of back and sides

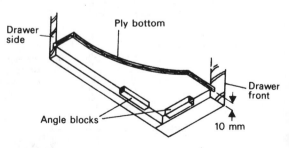

Drawer side

Ply bottom

Drawer front

Angle blocks

10 mm

Underside of drawer showing angle blocks supporting bottom

front and sides, but not the back. This groove is best kept up about 10 mm from the bottom edge of the drawer so that small angle blocks can be positioned on the underside to give support to the drawer bottom. It will be seen that the back of the drawer sits on the top side of the drawer bottom. It is also kept down about 10 mm from the upper edge of the sides to prevent the drawer contents from jamming as it is withdrawn and making the drawer action smoother and easier.

Materials used for drawer bottoms are plywood and hardboard. Except in quite small drawers plywood must be considered the more suitable material since the thickness can increase with the size of the drawer. 4 to 6 mm thicknesses are most commonly used.

The drawer bottom must be securely housed in the groove, the depth of which should be equal to the thickness of the bottom itself.

Drawer pulls are fixed to the front of the drawer and these can be flush fitting or of the projecting type.

Various methods of sliding arrangement are used, the most common being a hardwood runner which is screwed and glued to the unit carcass with a groove of the same size worked in the centre of the drawer side. A simple rebated method is also common and so, too, are the fibre composition slides which are in two parts. One part is let into the drawer side and the other fixed is to the framework.

When the drawer is pulled out it is normal for the front portion to drop due to the balance of the drawer. To overcome this tendency a *kicker* of 50 × 25 mm can be fixed above the side of the drawer as shown in the base unit on page 180.

Door construction
Most kitchen unit doors are flush pattern. Like

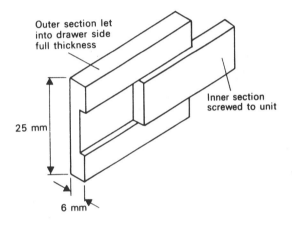

Outer section let
into drawer side
full thickness

Inner section
screwed to unit

25 mm

6 mm

Fibre drawer runner

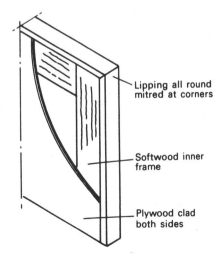

Lipping all round
mitred at corners

Softwood inner
frame

Plywood clad
both sides

Hollow core construction

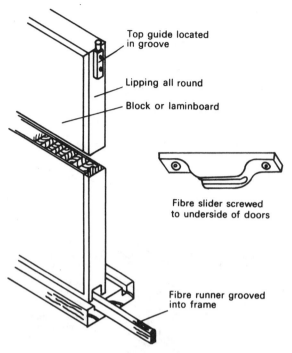

Top guide located in groove

Lipping all round

Block or laminboard

Fibre slider screwed to underside of doors

Fibre runner grooved into frame

Solid door on fibre sliding gear

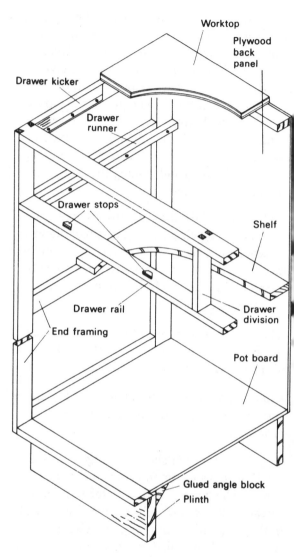

Worktop

Plywood back panel

Drawer kicker

Drawer runner

Drawer stops

Shelf

Drawer rail

Drawer division

End framing

Pot board

Glued angle block

Plinth

Base unit — general construction

their larger counterparts they can be of solid construction, e.g. blockboard, chipboard etc. or hollow. The hollow core type are usually of a softwood inner framework clad on both sides with plywood or hardboard. Whether solid or hollow core the doors should be lipped all round with 6 mm edging strip. This lipping must be mitred at the corners if it will be seen after completion.

Many doors are hung on hinges which have been specially designed for use on kitchen units. These hinges enable the door edge to abut the wall and yet still open to a maximum of 90°. Sliding doors are also popular for units in small kitchens where opening doors can be hazardous. Various methods of making the doors slide are adopted. Most common, nowadays, are the plastic or fibre types because these can provide smoother action. They are also capable of standing up to the wear on working parts more efficiently.

The better quality wall and floor units will have a back in them, even if it is only hardboard. A back panel will not only give a much better appearance to the inside of the unit, but will also make it a stronger, more rigid, unit. This is because the back panel can be fixed by pins or screws to the shelves, tops and sides as well as into the *potboard* at the bottom. A potboard is the bottom member of the base or floor unit, i.e. the shelf on which pots etc. may be stored.

Plinths are nearly always included on floor units because they enable the fitment to be scribed to the floor, thus taking up any discrepancy in the floor surface without interfering with the main body of the unit. Such plinths are sometimes quite separate from the unit itself, but can also be made integral to the framework as a whole.

Glass sliding doors are frequently fitted into nylon or fibre channels on kitchen wall units. The deeper channel is grooved into the *top* rail of the framework. This allows the glass to be pushed into the top first and then lowered safely into the bottom groove without falling out when resting on the bottom.

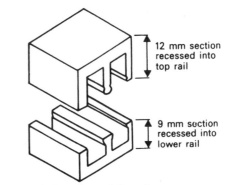

12 mm section recessed into top rail

9 mm section recessed into lower rail

Fibre track for glass sliding doors

Built-in wardrobes or cupboards

Fitments which are built-in on site form an increasing part of the carpenter and joiner's work these days. Such fitments include *wardrobes* (for the storage of clothing), *airing cupboards* (to house hot water tanks or cylinders where clothes are placed on slatted shelves to air) and other forms of storage cupboards.

These mainly consist of a frame constructed from material about 25 mm in thickness and between 50 and 75 mm in width. This frame should be mortised and tenoned throughout. Intermediate rails need only to be stub tenoned. The type of frame will depend on the requirements. For instance, most built-in wardrobes are designed with a pair of doors above and below the intermediate rail which separates the upper dead storage space (sometimes called the hat shelf) from the lower portion. The lower section should be at least sufficient for what is called 'long hanging'. This really means that the longest garments can hang on the tubular hanging rail without coming into contact with the bottom of the unit.

Wardrobes

Wardrobes are often built into a recess, which means that only the front frame is required. This is fixed to vertical battens on each side which must be fixed securely with screws and plugs. Much of the strength depends on this fixing. Occasionally, a return end may be necessary and this should be equal in width to the depth of the wardrobe. The return end should be framed up in similar materials to the front frame and clad with plywood over the whole surface. Where a plinth is incorporated in the design it is best for this to be of similar height to the skirting which is adjacent to it. In this way the line is carried round the room to give a pleasing effect.

Ordinary hinges are normally used on this type of fitment in conjunction with cupboard locks

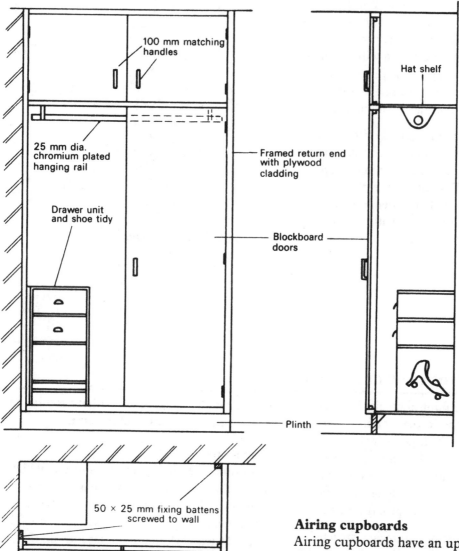

Built-in wardrobe

(if locking is desirable). Straight bolts are used on the inside of one door at the top and bottom. This door can only be opened after the first door has been unlocked.

Matching handles are fitted to the face of both the upper and lower pair of doors.

Airing cupboards

Airing cupboards have an upper and lower door, one of which gives access to the tank or cylinder while the other opens on to slatted shelving. Battens measuring 50 × 25 mm are often used for this shelving, with similar battens acting as bearers into which the slats can be screwed. There is no set size for the gap between the battens, but a widely used formula is to make it the same as the thickness of the battens being used, e.g. about 25 mm.

In turn, these slatted shelves are supported by bearers, normally 50 × 25 mm, which are fixed to the sides of the cupboard.

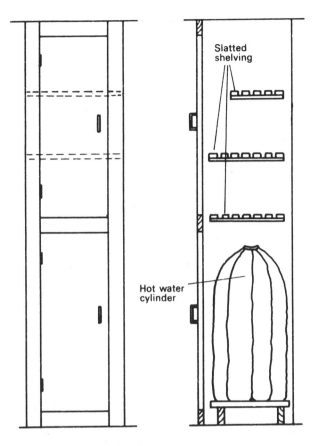

Slatted shelving

Hot water cylinder

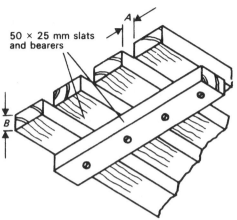

50 × 25 mm slats and bearers

Distance 'A' = distance 'B'

Airing cupboard

Since airing cupboards are seldom locked the doors are fitted with cupboard catches, usually of nylon, which are screwed to the inside of the door for neatness. Only a knob type handle is fixed to the outside face.

14 Panelling and casing

Panelling, which can be applied to walls and ceilings, is erected in one of two forms: framed panelling and direct cladding.

Framed panelling

This form has an outer frame which may include both vertical and horizontal intermediate rails. It can run from floor to ceiling, which is known as *full height*, or it can be *dado panelling* which covers the lower area of wall from the floor to the dado or chair rail — a height of approximately 1 m.

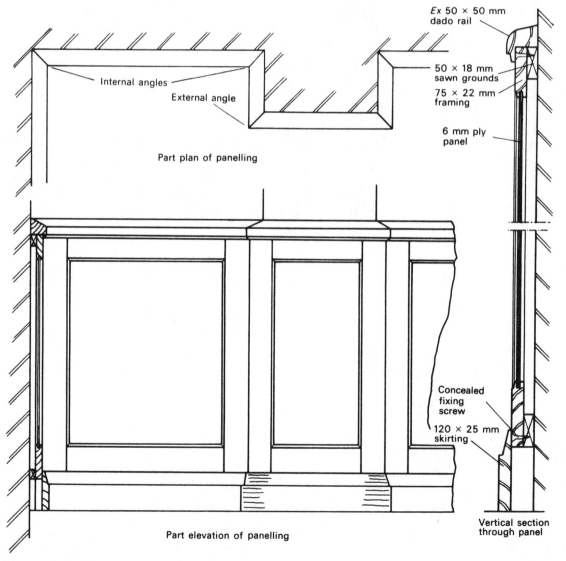

Ex 50 × 50 mm dado rail

Internal angles

External angle

50 × 18 mm sawn grounds

75 × 22 mm framing

6 mm ply panel

Part plan of panelling

Concealed fixing screw

120 × 25 mm skirting

Part elevation of panelling

Vertical section through panel

Framed panelling

If it is to be painted then softwood is adequate, but where a natural wood finish is desired then one of the attractive hardwoods would be more suitable. The framework would normally be around 75 × 22 mm with a wider bottom rail, perhaps 150 × 22 mm. The bottom rail width depends on whether a skirting board will be fixed to the face. If this is so, the bottom rail need only be wide enough to provide a margin of similar proportion to the rail width plus about 25 mm to run behind the upper edge of the skirting. The vertical members run to floor level to provide adequate backing for the skirting.

Panels are usually 6 mm thick plywood grooved into the framework. Mortises and tenons are used throughout the framework construction — those on outer members are haunched where appropriate and all intermediate tenons are stubbed.

Framed panelling is fixed to grounds which, in turn, must be securely and accurately fixed before the framework is applied.

The grounds are normally sawn softwood 50 mm wide and 18 mm thick. It is essential that the grounds are fixed at frequent intervals, preferably with rawlplugs and screws.
A straight edge or string line should be used to make certain that the grounds are not only

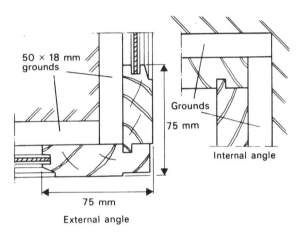

Framed dado height panelling

plumb (vertical) but also absolutely flat across the face of each ground. Packing pieces should be used behind the screw fixings if the wall surface is uneven. The packings are situated behind the screw to prevent them falling out and causing movement in the panelling if shrinkage should take place after fixing.

Direct cladding

The availability of a wide range of matchings in both hard and soft wood, as well as a large number of decorative faced sheet materials, has resulted in this form of panelling being widely practised.

For the work to be of a high standard, sawn

softwood grounds should be fixed securely and accurately as for framed panelling.

Matching boards of various design are widely used either vertically or horizontally. This type of cladding is often used on ceilings. The matchings are, in most cases, 100 mm wide and

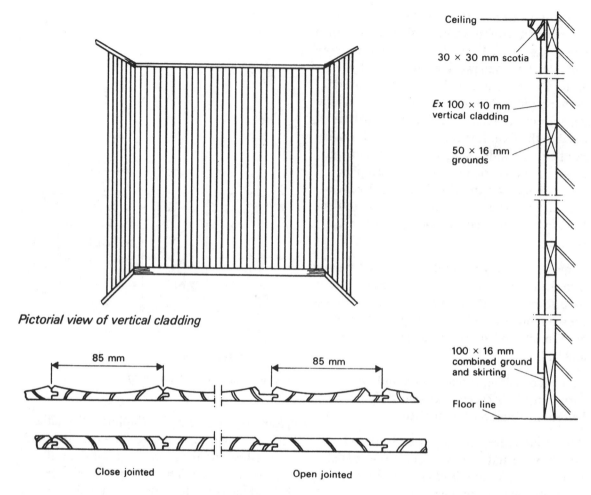

Pictorial view of vertical cladding

Ceiling

30 × 30 mm scotia

Ex 100 × 10 mm
vertical cladding

50 × 16 mm
grounds

100 × 16 mm
combined ground
and skirting

Floor line

85 mm

85 mm

Close jointed

Open jointed

Types of matching boards used for cladding

either 16 or 25 mm thick. Thinner boards are also used for panelling interiors, but care must be exercised to ensure the suitability of all such boarding with particular regard to moisture content. An equally wide range of 6 mm thick hardwood veneered plywood is also available for panelling walls, provided suitable grounds have been fixed prior to cladding.

The vertical and horizontal joints are normally covered with plain or moulded cover fillets. Whether the surface is panelled with matchings or sheet material the skirting is best formed as a recessed flat member. This gives toe space to prevent considerable damage being done by people accidentally kicking against the panelled areas.

Casing

This is the term given to the covering of pipes, beams or columns etc. which is frequently carried out by carpenters on-site. As for panelling, sawn softwood grounds form the base on which the covering or cladding material is fixed.

Materials for cladding vary with the type and requirements of the casing. Quite often matchings are used as in panelling or, alternatively, plywood can be used.

Beams

In some steel framed buildings British Steel Beams (BSB's) have to be *cradled* to enable plaster board to be fixed before being plastered. In casing of this kind a minimum of two thicknesses of plasterboard is required. The cradling can be framed in 50 × 25 mm softwood with corner halving joints formed at each angle.

Nogging pieces have to be cut accurately and driven hard into the web of the BSB. This is made easier by the fact that the rolled edge of the beam is tapered slightly, giving the nogging piece a wedge-like grip. Cradles should be fixed at not more than 600 mm spacings in order to make the cladding rigid.

Pipework

When pipework has to run from one floor to the floor above it is usually situated in the corner of the room. This permits the pipe or pipes to be cased neatly and avoids unsightly service pipes being on view.

A sub frame of 32 × 20 mm material framed together with halving joints should be made up with a return side of sufficient depth to clear the pipes which are to be cased. Battens of 50 × 25 mm should then be fixed securely and vertically to each wall so that the skeleton frame can be screwed to them. This now leaves only

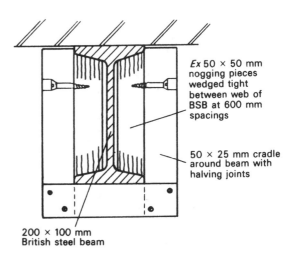

Ex 50 × 50 mm nogging pieces wedged tight between web of BSB at 600 mm spacings

50 × 25 mm cradle around beam with halving joints

200 × 100 mm British steel beam

Beam cradling

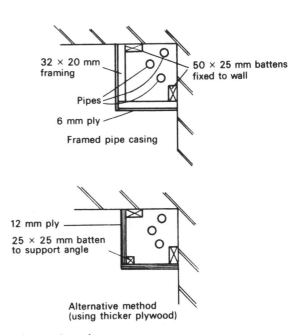

32 × 20 mm framing

50 × 25 mm battens fixed to wall

Pipes

6 mm ply

Framed pipe casing

12 mm ply

25 × 25 mm batten to support angle

Alternative method (using thicker plywood)

Pipework casing

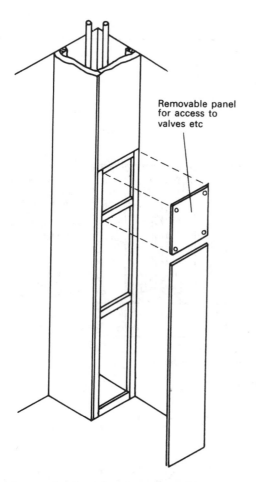

Part isometric view showing removable panel

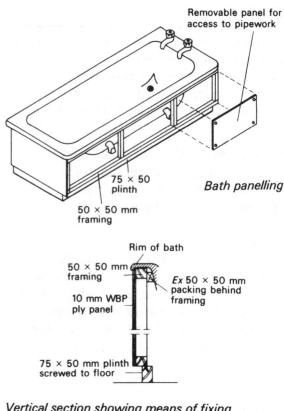

Bath panelling

Vertical section showing means of fixing framework

the cladding to be applied. Plywood is often used for this purpose because it can be prepared to receive paint, wallpaper or even ceramic wall tiles as a finish treatment.

Where pipes which are to be cased have a stopcock or valve, a removable access panel must be provided to enable future maintenance or repair to be carried out.

Baths

Baths often have to be panelled. The work must include removable access panels at the end of the bath where both the taps and waste are situated.

Any plywood used for this purpose should be of the WBP grade because of the condensation, heat and other extreme conditions which are frequently encountered in bathrooms.

The framework should be in 50 × 50 mm softwood which, when assembled, can be placed inside the rim of the bath and secured by packing pieces. A toe space should also be provided. This can be formed by screwing a length of planed 75 × 50 mm on the floor on to which the main frame can be seated and screwed.

The reason for such sturdy frames in bath panelling is, of course, the fact that in such a position the panel is likely to be frequently knocked or leaned against fairly heavily.

15 Temporary carpentry

A great deal of carpentry work is erected on site, only to be removed when the purpose for which it was built is completed. The three main categories of this *temporary carpentry* are:

1 *Formwork*
Commonly called 'shuttering', this is the temporary casework used for the casting of concrete '*in-situ*' (in position) and the moulds used for concrete components which are pre-cast (made out of position).

2 *Centring*
This is the name given to the work of providing temporary supports to brick or stone arches during construction.

3 *Shoring*
When a building becomes unsafe or is likely to suffer damage because of nearby works, a system of shoring can be used to 'prop' it up until it is made safe.

Formwork

With the number of concrete structures being built nowadays there is an enormous amount of formwork or shuttering which has to be carried out.

A great deal is also pre-cast off site in factories equipped to fabricate large and small sections of reinforced concrete sections for erection on site. Smaller components such as pre-cast lintels or coping stones are often made up on site, but not *in-situ*. For this type of work timber mould boxes are constructed which can be easily dismantled and refilled as many times as required. Some accuracy and attention to detail is necessary, especially where a good quality finish to the face of the concrete is specified.

Mould boxes
Take, for example, the construction of a box for a number of saddle-back coping stones required to surmount a brick wall with an overhang and drip on each side.

The material used must be able to withstand continual assembly and dismantling as well as having high resistance to water. Plywood of WBP quality and 19 mm thick is used in this case.

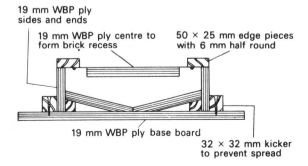

19 mm WBP ply sides and ends

19 mm WBP ply centre to form brick recess

50 × 25 mm edge pieces with 6 mm half round

19 mm WBP ply base board

32 × 32 mm kicker to prevent spread

Section through mould box showing general construction

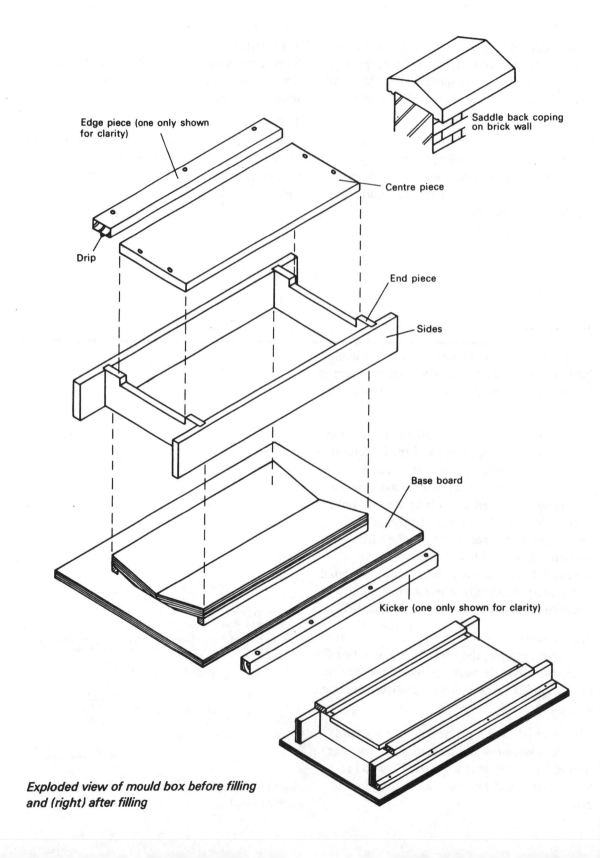

Edge piece (one only shown for clarity)

Centre piece

Saddle back coping on brick wall

Drip

End piece

Sides

Base board

Kicker (one only shown for clarity)

Exploded view of mould box before filling and (right) after filling

A *base board* should first be cut with about a 100 mm overhang on each side and at the ends. The sides and ends are cut to extend from the base board and to allow the full thickness of the coping stone. The saddle-back effect is formed by building up the bottom of the box with two splayed members. In addition, the end pieces are cut to receive a piece of ply that will form the groove which sits on to the brickwork. Drips are formed by pinning 6 mm half round on the two side members. Battens (about 32 × 32) are screwed throughout each long side to prevent spread or distortion when filling.

In some instances reinforced concrete is used for all parts of the main structure of the building including floors, walls, columns and beams. For economic reasons, therefore, it is necessary to re-use timber formwork repeatedly for the same operations. This is made easier by the regular and correct use of *release agents*. A release agent is a liquid which when applied to the inside faces of the formwork will prevent adhesion between the timber (or steel) and concrete surface. Adhesion can result in damage to the shuttering and produces a poor quality finish to the concrete.

Although a great deal of timber is used in this work steel is also quite widely used and it is not uncommon for a combination of timber and steel to be in use at the same time. When dismantling (striking) the formwork, whether steel or timber, care should be taken not to damage the material and it should be thoroughly cleaned and stored ready for re-use.

Lintel boxes

Apart from concrete structures it may also be necessary to erect formwork for reinforced concrete *lintels* which are to be formed *in-situ* on traditional brick-built structures.

Although this is, perhaps, a fairly straight forward example of formwork it must be borne

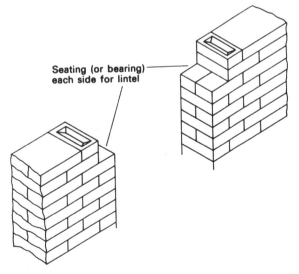

Seating (or bearing) each side for lintel

Brickwork opening prepared to receive formwork (or shuttering) for in-situ lintel

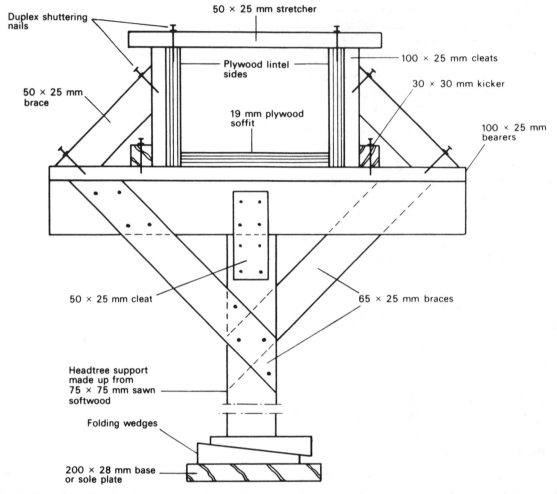

Vertical section showing formwork to in-situ *lintel*

in mind that the same principles apply to lintel boxes as apply to the more complex shuttering work. All shuttering must be constructed in such a way as to withstand the compound weight of wet concrete when it is poured and to hold the concrete in such form as required until it has cured.

The brickwork will be built up to the course which is in line with the top edge of finished concrete, making allowance for the appropriate bearing on each side of the opening. This bearing will vary according to the width of the opening, but a useful rule-of-thumb method of calculating it is to divide the clear opening width by eight. Only in special circumstances will it exceed 450 mm on each side.

All lintel boxes are made up of a soffit and two sides, normally cut from 19 mm ply of WBP quality. The soffit is cut accurately to the overall width of the lintel to enable the sides to be nailed to it on each brick face. Bearers of 100 × 25 mm softwood are fixed on the underside at not more than 600 mm centres. These bearers provide stiffening to the ply soffit and

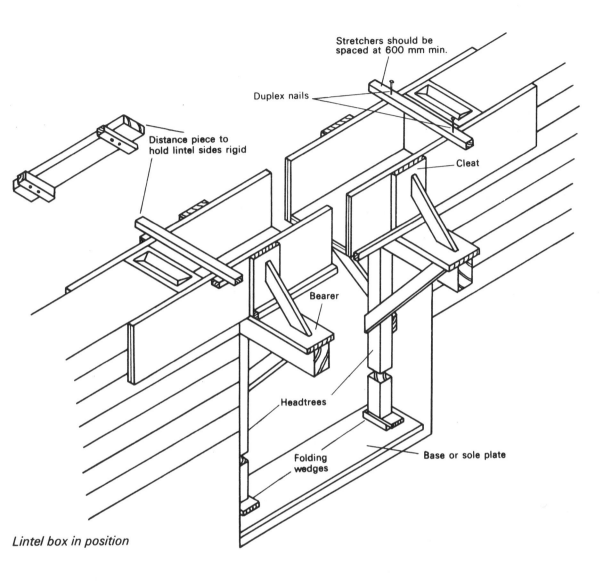

Stretchers should be
spaced at 600 mm min.

Duplex nails

Distance piece to
hold lintel sides rigid

Cleat

Bearer

Headtrees

Folding
wedges

Base or sole plate

Lintel box in position

act as a pad for the vertical supports of 100 ×
75 mm softwood. The over-run on each side
can be used for nailing bracing to the sides to
prevent deflection. Lintel box sides should be
allowed to run past the brickwork at each end
for a safe distance. Cleats of the same size as the
soffit bearers are nailed on the outside to line
up with the bearer position. This is to add
stiffness to the side and provides additional
fixing for the bracing, which can be 50 ×
25 mm sawn softwood.

Fixing the lintel box into position should begin

with the soffit being levelled in on its supports
with folding wedges positioned between the
soffit bearers and the supports to give any fine
adjustment that may be necessary before filling.

Folding wedges must be cut in pairs and with
only a slow drive (gradual taper) for accuracy.
The sides can then be fixed and braced. Metal
beam cramps are best for each end across the
brickwork, but *stretchers* of 50 × 25 mm sawn
softwood are nailed across the top of the box to
keep the structure parallel. Another method
using *distance pieces* with cleats is also shown.

Nailing in all formwork is best done with a double headed shuttering nail called a *duplex nail*. These are driven in until the lower head takes up, leaving the upper head projecting for easy withdrawal during the striking (dismantling) operation.

Wall forms

Formwork to walls is sometimes carried out in steel shuttering panels, but is also frequently constructed of 19 mm ply panels supported by 100 × 50 mm studs and braced. When the wall is of moderate size it will often be shuttered for and cast in one operation, but on larger walls a system of *climbing wall forms* is used.

Before either method is commenced, however, starter walls, which are commonly known as

'*kickers*', are cast in the exact position of the wall. These kickers are about 50 mm high and are positioned when the concrete floor slab is being poured. They provide an accurate start for the wall forms which are to follow.

The first stage in a climbing wall form would be to construct the form to be filled up to 1.2 m. *Rawl-ties* are used both as distance pieces and to ensure a uniform thickness for the wall. These ties are in three parts: a central section which remains encased in the concrete, and the two outer sections that are unscrewed after the concrete cures which also releases the plastic cone. It is into these threaded ties that the wall form is bolted for stage two of the operation.

The two most important aspects of wall

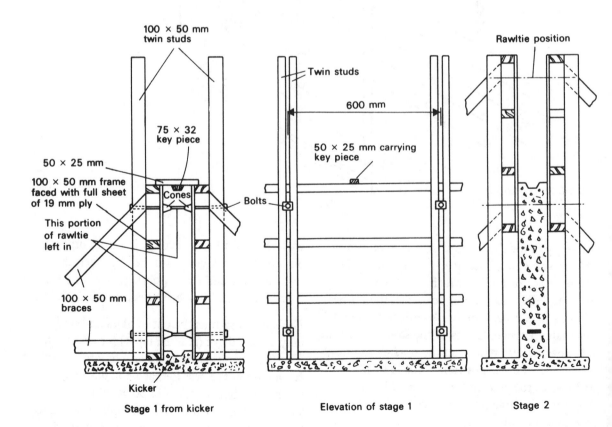

Climbing wall form

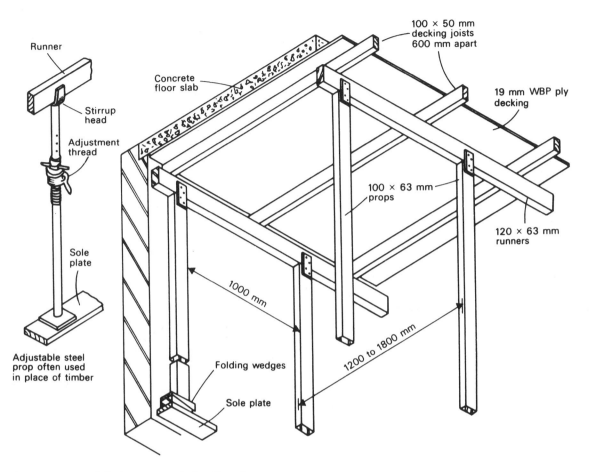

Arrangement of timber supports for floor form

forming are firstly that it is absolutely plumb (vertical) and secondly that it is sufficiently well braced and strutted to remain this way when being poured. A *key piece* of 75 × 32 sawn softwood is placed in the centre of the top surface on climbing wall forms to provide a key (groove) for the next section when it is cast.

Struts can be timber, or adjustable steel props which are widely used in all types of formwork.

Floor forms

Concrete floors of thicknesses ranging from 100 mm upwards must be supported during construction.

The decking used for this purpose is either 19 mm plywood of WBP grade or steel panels 600 × 600 mm. Plywood decking is supported by joists of 100 × 50 mm sawn softwood which, in turn, is supported by 150 × 63 mm runners. Timber or steel props are used to support the whole structure. Whether steel or timber the props should have a sole plate running throughout each row of props. Spacing of between 1200 and 1800 mm, depending on the thickness of floor slab, is sufficient. The props themselves should be spaced about 1 m apart.

Adjustable steel props have a screw adjustment

which enable any adjustment to be made. The props should have a stirrup attachment for the runner to sit into for safety.

Timber props are usually 63 × 100 mm sawn softwood and must have folding wedges at the foot of each for any adjustment required. Cleats of 100 × 25 mm about 300 mm long are nailed to each side of the props to retain the runner in position.

Where the distance between the bearing walls is 4 m or less a patent steel floor form can be used in conjuction with the steel panels mentioned earlier. These are in the form of telescopic steel beams which are triangular in section. They are placed at appropriate centres from wall to wall and do not require any other support from underneath.

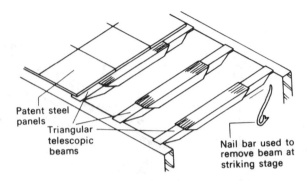

Patent steel floor forms for small spans

Columns and beams

The casings for beams and columns are often erected and filled in one operation, there being provision made at the top of a column case for the beam casing to run through.

As with wall forms, kickers (or starters) are cast during the pouring of the floor slab to give the column case a firm anchor when it is placed in position.

Column sides are usually cut from 19 mm thick shuttering quality plywood, although planed T and G flooring boards are also widely used. Plywood cases must be supported throughout their length and this is done with 100 × 500 mm vertical timbers at the corner of each face with intermediate rails being positioned to coincide with the *steel column clamp* positions.

Where T and G boards are used a different type of clamp called a *yoke* can be used. This consists of 75 × 50 mm pieces on each of two opposite sides bolted through on the other two faces with 15 mm diameter bolts with nuts and washers. Note that wedges are used between the

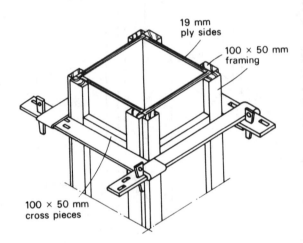

Steel column clamps on column case — made up as four sides before erection

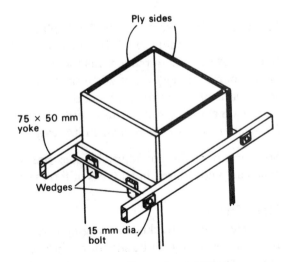

Timber yokes supporting column case

column case and the bolts to prevent distortion taking place.

Whichever method is used to clamp the case sides together, the intervals must decrease as they get nearer the bottom of the column. This brings the clamps or yokes closer together at the bottom to resist the *hydrostatic pressure* which builds up during the pouring of the concrete.

Beam sides and soffits are more often cut from 19 mm plywood although, once again, T and G boards are not uncommon. The soffit boards are, as in the lintel box, cut accurately and parallel to the overall width of finished beam.

In many cases the beam and the floor slab decking are erected at the same time. Because of this and the construction principles used there is ample support given to the plywood sides of the beam casing.

The soffit is supported by adjustable steel props or timber head-trees. The steel props also have the head-tree attachment when used in support of a beam soffit. Timber head-trees made from 100 × 75 mm have a cross-piece fixed by cleats across the top which, in turn, have 45° diagonal braces to support both ends, (shown on page 192). Adjustment to timber props is by folding wedges at the bottom between the prop and the sole plate. When the final adjustment has been made it is good practice to nail through the thin end of the top wedge into the thick end of the bottom wedge to prevent any accidental movement.

It will be noted that the floor decking bears upon the beam side. The edge of the floor decking in this position is splayed back to prevent the decking board from becoming trapped when the striking takes place.

In all formwork alignment (of columns etc.) levelling and plumbing (vertically) are of paramount importance and time must be spent

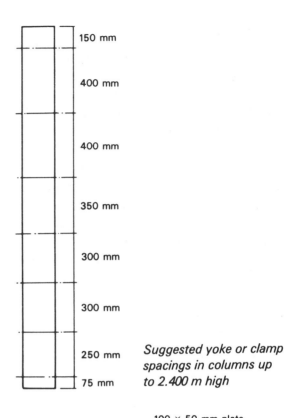

Suggested yoke or clamp spacings in columns up to 2.400 m high

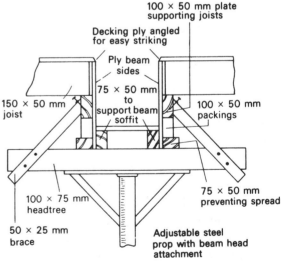

Detail of beam casing

on checking them before concrete is poured. Striking, too, must be carried out sensibly and carefully in the interest of safety and for maximum re-use of timbers and forms.

Centring

When a brick or stone arch is to be built, the carpenter must provide support for it. This support must be strong enough to hold the voussoirs (bricks or stones forming the arch) in position and also be exactly to the required shape of the arch. This is called a *centre*.

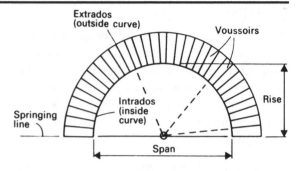

Semi-circular brick arch

Arches can be any desired shape although the majority are *semi-circular* (half circle), *segmental* (less than half circle) or *elliptical*. Those arches whose shape is a part of a circle are reasonably easy to set-out, but elliptical arches pose a different problem and have a number of ways by which they can be set-out.

Probably the most simple method of setting-out an ellipse is what is known as the *trammel method*. The longest axis of the ellipse is called the *major axis* and the shortest is called the *minor axis*. The trammel (a small flat batten) is marked with half the major and half the minor axis on it. With the minor axis mark moving along the major axis and the major axis mark moving along the minor axis the points are marked at frequent intervals until the full curve has been drawn.

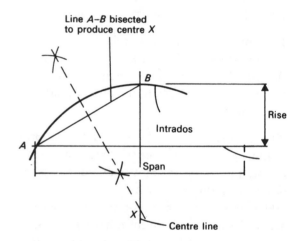

Setting-out a segmental arch

The line or level from which the curve of the arch commences its shape is called the *springing line*. In elliptical arches the springing line is usually on the major axis.

When setting-out any arch centre it is important to pay particular attention to the *span*, which is the clear opening size between brick or stone jambs, and the *rise* which is the vertical height in the centre of the arch measured from the springing line.

If the arch is of only slight curvature with only a small rise which does not warrant a built up arch centre, it can be supported with a *turning piece*. This is cut from a solid piece of timber, taking care to shape the curve accurately to conform to the shape of the arch.

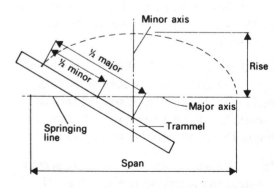

Using the trammel to set-out an elliptical arch

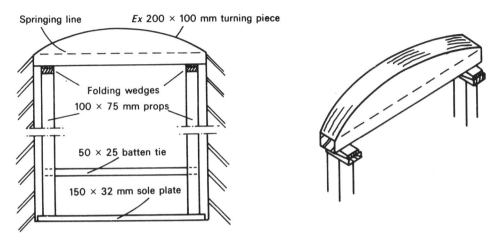

Solid turning piece for segmental arch of low rise

Turning pieces and small arch centres are supported by props of 100 × 75 mm sawn softwood which stand on 150 × 32 mm sole plates. Folding wedges are used for final adjustment, but in arch centering they are placed at the top of the supports just below the springing line.

Arch centres up to 1 m span are often made up in a simple fashion with 19 mm WBP plywood. Ribs are cut to the shape required and each truss consists of two layers of ply nailed together for greater strength.

The thickness of the wall which is to be arched will determine how far apart the two rib frames will be, but in all cases *laggings* have to be nailed to the outer curve (*extrados*) of the arch. In small arch centres the laggings will often be made up of two layers of thin ply (exterior grade) nailed to the outside curve of the centre, whereas in centres for larger spans the laggings will be in batten form. In addition to providing a surface for the arch to sit on during construction, the laggings also act as a brace between the two trusses.

Always remember to allow for the thickness of the laggings when setting out the curve of the centre.

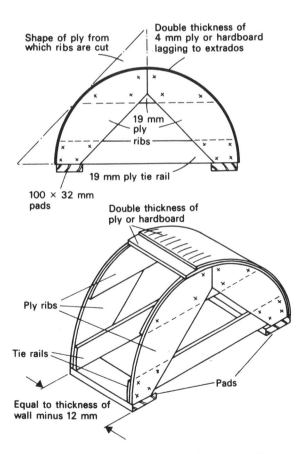

Suitable centre for segmental arches of small span

Laggings should be kept back about 6 mm from each face of the wall to avoid interference with bricklayer's lines or levels.

As the span increases the design of the truss changes so that its strength will increase to take the extra loading which will be placed upon it. These larger trusses are also built up with two layers of board each 25 mm thick. The *ribs* are those members which are shaped on the outside (extrados) and the supporting members which generate from the centre of the curve are called *struts*. Across the springing line is a member called the *tie* which must always be in one piece because it prevents the spread of the centre when under load. All the joints are simple butt joints which must be cut accurately and each truss is assembled in such a way as to ensure that all joints are staggered for maximum strength. To give further rigidity to the centre a diagonal brace can be nailed between the struts in the depth.

Laggings for this size of arch are sawn softwood battens 19 mm thick. The width is dictated by the severity of the curve because laggings must not be so wide that they rock on the arch curve.

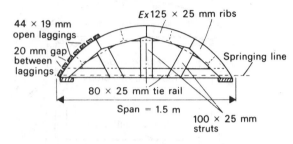

Trussed segmental arch centre

In gauged brick or stone arches the laggings should be nailed side by side in a continuous line, this is called *close lagging*. In other arches *open lagging* with spaces of about 20 mm between battens is adequate. Laggings should overhang the truss by 12 mm on both sides.

Wider arch spans require a pair of support posts at each end braced together to prevent spread, with folding wedges for adjustment and easing. A segmental truss is shown in elevation for a span of 1.5 m together with the part elevation and section of a semi-circular trussed centre suitable for a span of 3 m.

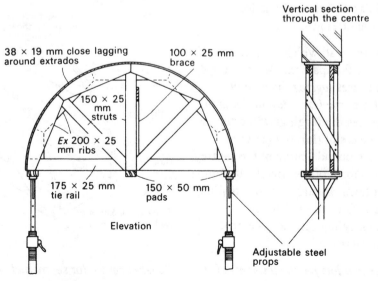

Suitable trussed centre for spans up to 3 m

Easing

When the brick or stone arch is set the centre should be *eased* before dismantling. This means that the folding wedges are knocked back to allow a fractional drop (about 4 mm) in the arch centre. This takes up any settlement which may occur in the newly built arch. The centre should remain in the eased position for a day or two.

Shoring

Two main systems of shoring are used fairly widely: dead shoring and raking shoring. The principles of each should be studied closely and followed whenever shoring is required to a building.

Dead shoring

So called because this type is used to carry the dead load of the brickwork etc. when an opening is to be formed or when an existing opening needs to be widened.

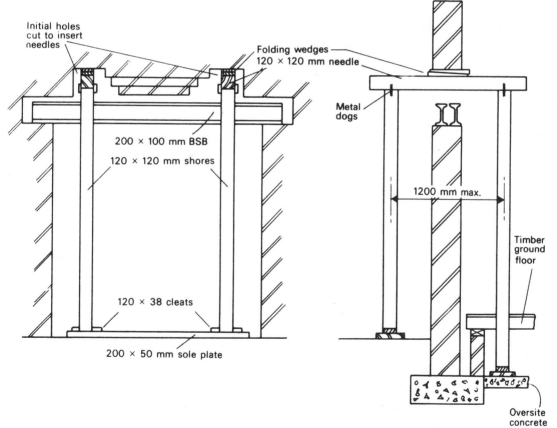

Dead shoring used to enlarge opening and insert beams

Vertical section showing inner shore taken through timber floor to bear on oversite concrete

Timbers of square section are positioned above the line of the lintel or BSB. These are called *needles* because they pass through the wall. They are supported by timbers of similar size called *shores*. The shores must be stood upon sole plates which, in turn, must stand on firm ground — preferably concrete. Under no circumstances should shores be placed on suspended timber floors.

Shores should not normally exceed 1200 mm centre to centre, but the size of the opening and the amount of load to be supported above will dictate their size and spacing. Cleats are nailed each side of the shore at sole plate level to prevent it becoming dislodged, while at the upper end metal dogs are used to secure the needle to the shore.

Whatever number of shores and needles are used they should be linked together by the sole plates and bracing where necessary. The main consideration for the positioning of the shores is that they should in no way interfere with placing the lintel or beam into position. They should also give the bricklayer room in which to work.

Folding wedges can be used to take up any discrepancy where the brickwork being supported comes into contact with the needles. Where there are window openings above the dead shores they should be *strutted* in the manner shown. This will prevent any re-directed forces above the shoring from causing distortion or fracture to brickwork around the openings.

Raking shores
These are used for very different reasons from the dead shore, although there may be occasions when the two are used in conjunction with each other.

Raking shores are designed to give external support to a building which may have become

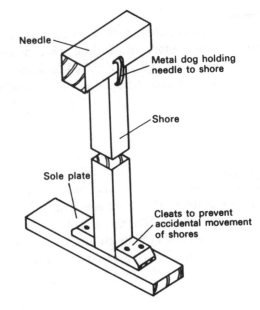

Position of cleats and dogs

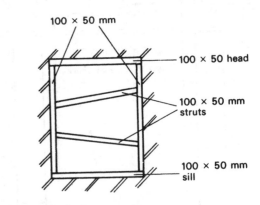

Strutted window opening

in danger of collapse. They are more rarely used when the building is adjacent to a site where construction work is about to be undertaken which could cause serious movement or vibration and the shoring is erected for precautionary reasons.

The system may consist of one or more rakers depending upon how many floors the building possesses. Those with a number of floors to support are known by the number of shores used, e.g. single raking shore, double raking shore etc. A single raking shore consists of a *wall piece* of 225 × 75 mm secured flat against the wall with metal wall hooks against which the shore is positioned. At the head of the 150 × 150 mm *raking shore* a needle of 100 × 100 mm passes through the wall plate about halfway into the wall thickness.

The angle of the shore should be about 60° and in no case should it exceed 70°. Immediately above the needle a cleat is fixed and housed with a splayed housing into the wall plate, this will support the needle in taking the thrust. Particular attention must be paid to the position of the raking shores in relation to the floors. It will be seen from the illustration that the centre line of the raker picks up the corner of the wall plate carrying the joists. Where the joists are parallel to the wall being supported the centre line of the shore intersects with the centre line of the floor thickness in the centre of the wall thickness. This intersecting point is called the *node point*.

The foot of the shore is seated on to a sole plate of 225 × 75 mm which should be placed on very firm soil or weak concrete mix. An angle of 80° between the raking shore and the sole plate will assist in allowing the raker to be eased into position (tightening as it goes) without undue hammering or force which would certainly be unwise with a wall already becoming unsafe. A notch is formed in the foot

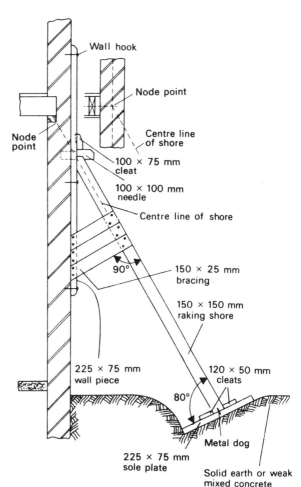

Raking shore

of the raker to enable a nail or crow-bar to be used for easing the shore into position. Having obtained maximum tightness on the raker a cleat should be nailed on top of the sole plate hard against the foot of the shore to prevent any loosening. In addition, metal dogs are used to secure the shore to the sole plate.

Bracing is used to tie the shore to the wall plate and give more strength to the structure. Such bracing is usually 150 × 25 and should be fixed at right angles to the rake of the shore.

This type of shore may well be used in conjunction with dead shoring if the opening to be formed in the wall of the building is of such proportions that the remainder of the wall is liable to be disturbed.

All shoring systems should be eased in the same way as arch centres before being totally dismantled. Failure to do this may result in sudden movement which could cause further damage to the structure.

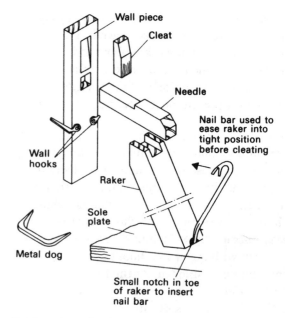

Raking shore foot

16 Timber frame buildings

Two types of timber frame building are erected by carpenters on site. Both types are reasonably straightforward, provided some knowledge of their construction is known. Those referred to are:

1 *Site huts*
These are erected on site at the commencement of work and are used to house offices, stores, toilets and mess-rooms for the duration of the job.

2 *Timber frame houses*
This type of dwelling is still fairly widely used because it provides a warm, comfortable home which does not require heavy plant and lifting equipment to carry out construction on site.

Site huts

Softwood framing is used in the prefabrication of these huts, 75 × 50 mm being a very common size. Because of the fact that they have to be transported from site to site the panels are kept to a size which is easy to handle. Construction is of the simplest kind and in some cases the panels are butt jointed only. However, the better quality huts will have *open (slot) mortise and tenons* at each angle. Openings for doors and windows are formed within the studded frame where required, but both doors and windows are of simple design and construction. Ledged and braced doors are often used and a sash of simple construction is top hung into a simple lining which surrounds the opening.

Side and end panels are stood on and bolted to the floor sections. The floor sections should be laid into position first unless the floor is to be concrete, in which case they stand directly upon the concrete. The corners of the panels are bolted together with coach bolts and washers to provide a rigid structure.

The site must be levelled off over the whole area of the hut. It may even be necessary to provide timber sleepers from which to commence the erection.

Roof sections are either single or double pitch, but whichever is used the angle of pitch is always low, seldom being more than 22 ½°. These are bolted to the top rail of the wall panels. The height to the eaves rarely exceeds 2 m, which permits the use of standard height doors.

If some site personnel are to work in the hut all the time, it is sometimes lined internally to give extra comfort. The lining in most cases is 3 mm hardboard, but before it is fixed into position insulation quilt is placed in the cavity between the internal cladding and the inner lining.

External cladding
External cladding can take a number of forms. Simple feather edge boards nailed in a horizontal position are very common, as are shiplap boards and horizontally fixed T and G vee jointed match boards. In this type of construction a layer of felt is sometimes fixed to the framework immediately under the cladding to act as a water barrier.

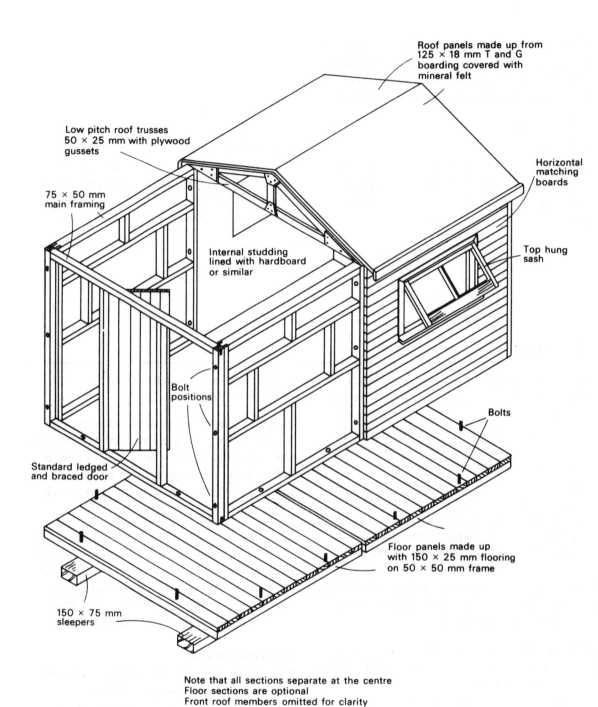

Roof panels made up from
125 × 18 mm T and G
boarding covered with
mineral felt

Low pitch roof trusses
50 × 25 mm with plywood
gussets

Horizontal
matching
boards

75 × 50 mm
main framing

Internal studding
lined with hardboard
or similar

Top hung
sash

Bolt
positions

Bolts

Standard ledged
and braced door

Floor panels made up
with 150 × 25 mm flooring
on 50 × 50 mm frame

150 × 75 mm
sleepers

Note that all sections separate at the centre
Floor sections are optional
Front roof members omitted for clarity

Typical site hut construction

Timber frame houses

There are many reasons why timber frame housing is still popular. They can be erected very swiftly on site without expensive equipment and are easy to insulate for heat and sound, thereby cutting costs and increasing comfort. However, these houses are not exempt from the Building Regulation requirements — like all other building works they must fully satisfy the regulations.

The houses can be single storey (bungalow) or have ground and first floors. They can be detached or semi-detached.

The two principal methods of construction are:

Balloon framing

In this method it will be seen that the external studs run from ground level up to under eaves and, with the roof, form a continuous external

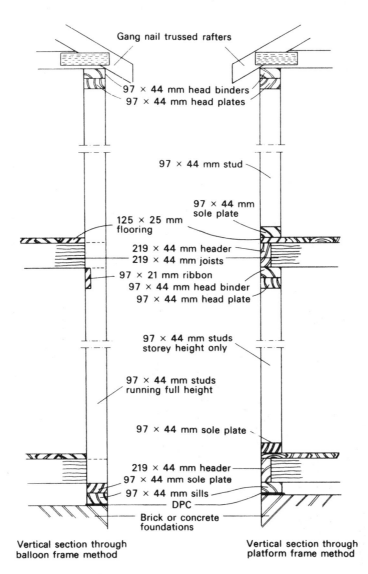

Vertical section through
balloon frame method

Vertical section through
platform frame method

Methods of constructing two storey timber framed houses

line rather like a balloon. A *head binder* and *sill* are used above and below the stud frame. Carrying the first floor joists is a member called the *ribbon* which is 97 × 21 mm PAR (planed all round) and recessed full thickness into the stud framing.

Platform framing

This method is more popular nowadays than balloon framing. There may be several reasons for this, among them being the comparative simplicity of on-site work. The fact that wall frames are only storey height greatly reduces the amount of plant or mechanical apparatus necessary to erect the frames.

It will be seen that the effect of storey height framing is to 'sandwich' the floor between the storeys and this results in a very rigid structure.

Unlike structural timbers in traditional buildings these timbers are planed all round before use, and in addition, all timber used must be stress graded: it must be mechanically or visually tested for strength. Two grades are used: GS (general structural) and SS (special structural). Wall frames are GS grade, whilst lintels and joists are SS grade.

Wall frames are made up from 97 × 44 planed softwood and should not exceed 3.600 m in length, the height being a standard 2.365 m.

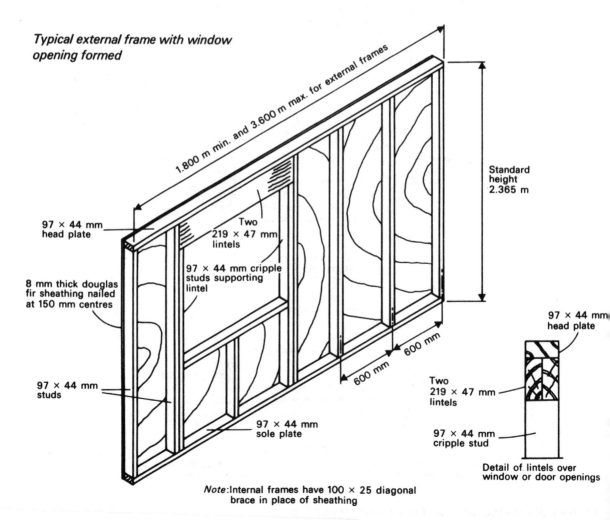

Typical external frame with window opening formed

1.800 m min. and 3.600 m max. for external frames

Standard height 2.365 m

97 × 44 mm head plate

Two 219 × 47 mm lintels

97 × 44 mm cripple studs supporting lintel

8 mm thick douglas fir sheathing nailed at 150 mm centres

97 × 44 mm studs

600 mm

600 mm

600 mm

97 × 44 mm head plate

Two 219 × 47 mm lintels

97 × 44 mm cripple stud

97 × 44 mm sole plate

Detail of lintels over window or door openings

Note: Internal frames have 100 × 25 diagonal brace in place of sheathing

Studs are placed at 600 mm centres in order to make maximum use of sheet materials. All frame parts are butt jointed.

Window and door openings are formed where necessary. Each opening over 1.500 m has a lintel above supported by what are called *cripple studs*. Lintels are 219 × 47 mm and are paired for each opening. Openings smaller than 1.500 m are framed up in the normal way without lintels. However, no opening in a wall frame should exceed 2.100 m.

All exterior wall frames are clad with 8 mm thick Douglas fir plywood sheathing which must be WBP grade. The exterior wall frames are held together by a 97 × 44 mm *head binder* which is nailed into the top edge of the frame.

Interior wall frames which form the partitions within the house are made up in the same way as the exterior frames, with the exception of plywood sheathing which is not used. Some on-site use of 97 × 44 mm studs for the fixing of internal partitions will be necessary.

Trussed rafters of the kind described in Chapter 9 are used for these houses and are between 22 ½ ° and 40° pitch. The rafters are spaced at 600 mm centres so that they will be positioned directly above the wall frame studs as well as the floor joists below. The trussed rafters are seated upon the head binders on the front and rear external walls.

Bungalow type buildings

The sequence for the erection of a bungalow type timber frame building after the concrete foundations up to DPC (Damp Proof Course) level would be as follows:

1 *Lay base plates*
These will give correct outline of building for the joists and floor decking to be permanently fixed.

2 *Erect, plumb and brace external and internal wall panels*
This will allow the head binder to be fixed thoughout, making certain that joints of head binders and wall panels are staggered.

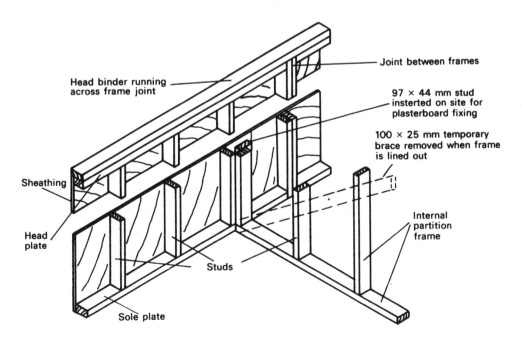

Head binder running across frame joint

Joint between frames

97 × 44 mm stud insterted on site for plasterboard fixing

100 × 25 mm temporary brace removed when frame is lined out

Sheathing

Internal partition frame

Head plate

Studs

Sole plate

View of external frame with internal frame abutting

3 *Erect and brace trussed rafters*

The roof is then covered with sarking felt to allow work to continue inside the building.

4 *Doors and windows*

All door and window frames of standard joinery pattern are fixed into position using DPC materials to ensure a water tight joint all round.

In order to keep out the weather, all the outside faces of the sheathing are clad with a special building paper called *breather paper* which acts as a moisture barrier. Horizontal joints should have 100 mm lap for safety and vertical joints 150 mm lap.

The void between the studs is filled with insulation quilt. A vapour barrier is then fixed to the inside face of the internal wall framed under the plaster board. The vapour barrier is usually polyethylene sheeting of 250 gauge thickness. As a double precaution foil backed plaster boards should also be used.

External cladding

External cladding can consist of a single skin of face brickwork if desired. Very often, however, a lightweight cladding of timber such as *shiplap*

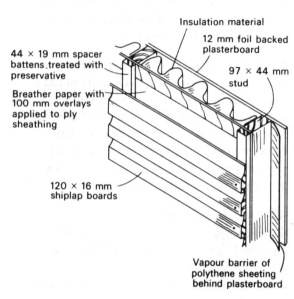

Horizontal shiplap external cladding

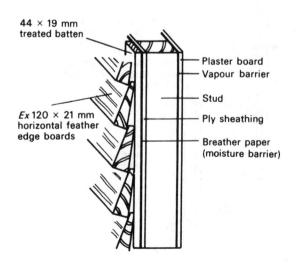

Feather edge boards as an external cladding

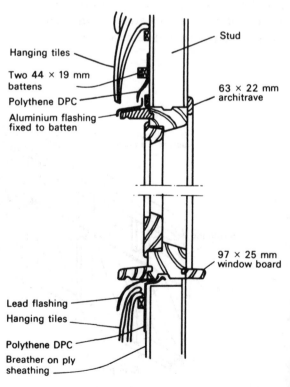

Detail of weathering around standard window

or *feather edge* matchings is used. It will be readily appreciated that where this type of cladding is used it would be best if it is either Western Red Cedar, because of its resistance to decay or, if another softwood is used, it should be treated with a preservative. Either of these matchings must have a minimum thickness of 16 mm. Shiplap (or other) claddings are normally fixed horizontally on treated battens 44 × 19 mm fixed on to the face of the breather paper.

A two storey house of this kind will embody all the principles which have been described in this single storey example.

Questions

1   Outline some of the reasons for the continuing popularity of timber frame houses.

2   List and describe the two principle methods of timber frame construction.

3   Where is the *head binder* to be found and what is its function?

4   What is the purpose of *cripple studs*?

5   List the sequence of operation when erecting a timber framed building.

6   Where are the moisture and vapour barriers inserted and what are their functions?

17 Ironmongery

Ironmongery is the collective term used to describe the many items made from metal which are and always will be associated with the work of the carpenter and joiner. These not only include the locks and hinges that are fitted by the carpenter, but also the means of fixing them, such as screws, nails etc.

There is insufficient room to study the full list here and so this section will confine itself to the more common forms of ironmongery.

Nails

Too many woodworkers have thought that the study of nails and nailing is unworthy of the time spent on it. This view is unfortunate in the extreme because many jobs have been marred by using an unsuitable nail for the task in hand or by an incorrect method of nailing.

When a nail is driven in the fibres of the timber are forced apart, which will tend to cause some splitting of the grain. Depending on the position of the nail it may, therefore, require a small pilot hole prior to the insertion of the nail.

We saw in fixing the herringbone strutting in a single floor, how a small saw cut in the edge reduced the tendency to split. However, it is only when nailing near the end or edge of boards that such precautions need to be taken.

The nails and pins which are most commonly used in carpentry and joinery are:

1 French wire nail
2 Lost head nail
3 Clout nail
4 Oval brads
5 Panel pins
6 Veneer pins
7 Hardboard pins
8 Cut clasp nails
9 Floor brads

French wire nail
Extensively used in carcassing and other forms of framework. They are made from stout wire,

French wire nail

hence the name, with a flat round head which is roughly twice the diameter of the stem. Sizes are available from 25 mm, in 12 mm sizes, up to 150 mm. Rafters, joists etc. are nearly always fixed with this nail.

Lost head nail

Also made from wire, but with a much smaller head which enables it to be punched in below the surface. These are used on finishing work where the strength of the wire nail is maintained without having the unsightly head showing.

Lost head nail

Clout nail

Another wire nail, but this time with an extra large round head designed primarily for board fixing to stop the head pulling through the board. They are quite often galvanised to prevent rusting when used to fix ceiling plasterboard or wallboard. When plastered over they will not stain through the finished plaster. Smaller clout nails, about 20 mm long, are used to attach sash cords to sliding sashes.

Clout nail

Oval brads

As the name implies, these are made from oval wire with a head so designed that it can be punched in and filled to give a high class finish. Ovals are strictly for finishings such as architraves, mouldings etc., where they achieve a very strong fixing. Care must be taken to ensure that the widest part of the nail is placed in the same direction as the grain to get maximum hold and reduce the possibility of splitting.

Oval brad

Panel pins

These are more often to be found in the joiners' shop where they are used for pinning small mouldings, beads and plywood in conjunction with gluing. Made from wire with a tapered head, they are available in lengths from as little as 10 mm upwards.

Panel pin

Veneer pins

Identical to panel pins in every way except that they are made from a thinner gauge wire.

Hardboard pins

These were developed with the aim of fixing hardboard sheets without needing to be punched in. They are small and square with a double taper on the head to enable them to be hammered just below the surface. These pins have a copper coating to prevent corrosion from staining the hardboard surface.

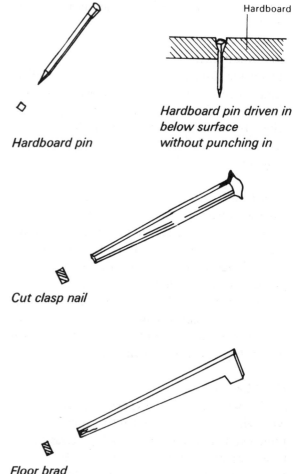

Hardboard pin

Hardboard pin driven in below surface without punching in

Cut clasp nail

Very often called simply a 'cut nail'. These are pressed from sheets of steel and are used for nailing into brick or blockwork when fixing skirting or other fixtures. They are capable, because of the tapered sides, of gripping the surrounding material and giving a very firm hold. Sizes range from 38 mm to 100 mm.

Cut clasp nail

Floor brad

This is similar to the cut nail insofar as it is pressed from a sheet of steel. They are of slightly different design and of one length only (55 mm). Intended for the nailing of flooring, the head is shaped so that it will not allow the board to rise after fixing and, in addition, will go just below the surface if fully hammered down.

Floor brad

Screws

Screws provide an excellent means of fixing, whether it is simply wood to wood, or wood to another base using plugs into which to turn the screw. They are identified by their length, gauge number and type of head. Screws are made of various metals such as brass, aluminium and steel (the most common), and in various finishes, e.g. black japanned, chromium plated etc. The points from which the length of a screw is determined can be seen from the illustration.

The method of driving the screw in is by means of a *standard slot* in the head, a *Phillips cross head* or the *Pozidriv star head* which is a refinement of the cross head. Most screws now available can be obtained with any of these slots to suit most requirements.

Standard slot

Phillips cross head

Pozidriv star head

Flat head countersunk

This type is used far more than any other. A clearance hole of slightly larger diameter than the shank must be drilled through one of the pieces in the desired position. In some softwoods the screw will pull itself in until the head is flush or just below the surface, but it is best to countersink the hole with a countersink bit and be sure that the screw will finish with a snug fit. A pilot hole of smaller diameter than the screw thread should also be made. This can be made with a bradawl in softwood, but a drill in the wheelbrace should be used for hardwood.

Round head

These screws have a decorative as well as a functional role. For this reason most round head screws are finished in brass, chromium plate or are black japanned, which is an enamel coating over the whole screw.

Raised head

Something of each of the above screws is featured in the raised head. It is countersunk on the underside with a segmental round on the top side. Once again they are decorative and are often used in conjunction with *screw cups* if it is thought that the screw will have to be taken out for panel removal etc. *Screw sockets* are also used for this purpose, but whereas cups are entirely face fixed, sockets are countersunk into the timber.

Dowel screw

A double ended screw of varying length and gauge. Dowel screws provide a useful fixture with glue where two pieces of timber are required to be fixed end to end.

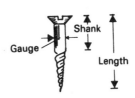
Flat head countersunk screw

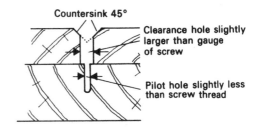
Preparation for screwing two pieces together

Length measured from point shown for round and raised head screws

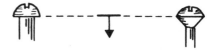

Round-head screw   *Raised head screw*

Screw cup (face fixed)

Screw socket

Dowel screw

Self tapping

Used for fixing to thin metal sheet or plastic where the thread (which goes full length) will cut itself into the material and secure a fixing.

Self tapping screw

Chipboard screws

Once again, the thread is the full length of the screw and is capable of getting a firm grip on chipboard.

Chipboard screw

Coach screw

This is a hexagon headed screw used for carcassing and heavy framework. A spanner and not a screwdriver is used and washers are necessary. Sizes vary to suit most requirements.

Coach screw

Domed head

Sometimes called 'mirror screws' because they are often used for screwing mirrors to the wall etc. They are useful for any panel or fitting that needs to be removed from time to time. The screw is usually brass with a threaded hole in the centre of the flat top into which a removable dome, having a threaded bolt attached, is screwed. The dome can be polished brass, chromium plated or any other finish which is desired.

Domed head screw

Gauge number

A useful 'rule of thumb' method of determining the gauge number of a screw is to measure the number of 1/16ths of an inch across the head of the screw. This number is then doubled and two is subtracted to give the gauge number. For example, a number 8 screw will measure 5/16ths of an inch across the head. The 5 is doubled, to make 10, and reduced by 2 to give the answer 8 — the gauge of the screw.

Hinges

Many types and sizes of hinge are made. The carpenter and joiner has to know which type is most suitable for the work to be done and also the correct method of fixing.

Most hinges are made of one flap on each side of a pivot which is called the knuckle. They are referred to in size, type of material, e.g. brass or steel etc., and type.

Those which the woodworker is likely to meet fairly regularly are:

1 Butt hinges
2 Loose pin butts
3 Lift-off hinges
4 Rising butts
5 Parliament hinge

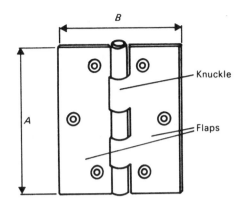

Butt hinge — fully open

Butt hinges

These are the ordinary hinges most commonly used. They are mainly steel, but are also available in cast iron, brass, aluminium or in tough nylon plastic. There is a range of sizes from about 25 mm up to around 150 mm. The size of this type of hinge is identified firstly by the length of the knuckle 'A' and secondly, by the width 'B', across both flaps when fully opened. The illustration shows a butt hinge fully open.

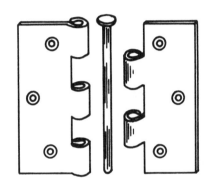

Loose pin butt

Loose pin butts

This is similar to the plain hinge, with the exception that the pin is removable when required so that the door can be taken off without removing the screws.

Lift-off hinges

This is another type which enables the door to be removed without taking the screws out. Care must be taken to ensure that the flap having the pin fixed to it is fitted and screwed to the frame or lining while the flap with the sleeve is screwed to the door.

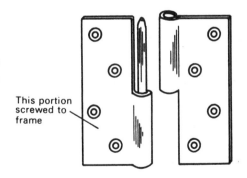

Lift-off hinge

Rising butts

This is a two part hinge which will raise the door by about 10 mm, enabling a very tight joint when the door is closed. Because of the design the lifting action commences as soon as the door starts to open. Rising butts are right or left handed to suit the opening of the door. When deciding which hand is required the door should be viewed from the side where the knuckles will be seen. If the knuckle will be on the left hand jamb it means that a *right-handed* pair of butts will be required. *Left-handed* butts are required for a door whose butts will be on the right hand side.

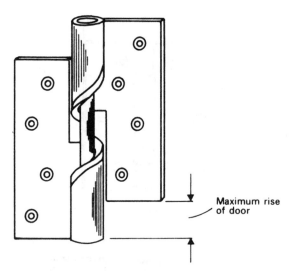

Rising butt hinge (left-hand)

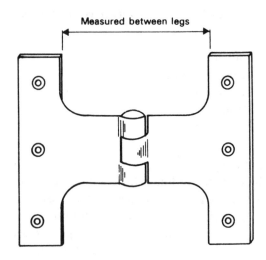

Parliament hinge

Parliament hinges

These hinges are used when it is desirable for a door or window to open with a gap between it and the frame to which it is hung. This is useful for windows at high level where it enables the outside of the window to be cleaned easily from the inside. This type of hinge is measured between the legs when fully open.

Locks and latches

The following are a representative selection of those locks and latches which are in constant use:

1 Mortise locks and latches
2 Rim locks
3 Cylinder rim night latches
4 Knobsets

Mortise locks

Most mortise locks are of the upright type, which means that the spindle and escutcheon (key hole) are immediately above one another, as shown on page 94. This enables the lock to be shorter in length, which means that less of the door stile is cut away to accommodate it. They are seldom more than 100 mm long and, for narrow stiles, can be obtained much shorter. The majority are reversible which makes them suitable for doors hung on either side.

Mortise latches

These are used mainly on internal doors where no locking mechanism is required. The furniture used on mortise locks and latches also varies greatly in both design and type of finish. However, the long plate type is very widely used because it includes both the handle (usually lever pattern) and escutcheon where appropriate. Finishes include bronze, brass and aluminium.

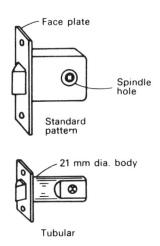

Mortise latches

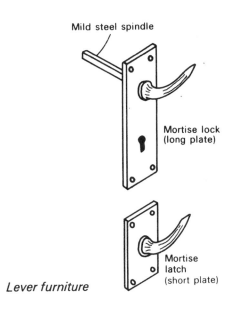

Lever furniture

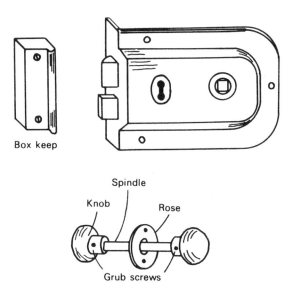

Rim lock and rim furniture

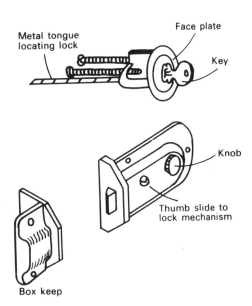

Cylinder rim night latch

Rim locks

Used mainly on out-buildings and sheds etc., these locks are entirely face fixed. They are usually 150 mm long, but other sizes are also made. Furniture for these locks is often knob pattern with separate escutcheon plates.

Cylinder rim night latch

These are primarily for situations where the door is to be kept locked, such as front entrance doors. These locks are in two parts. The cylinder, which is key operated, is fixed to the face and runs through the door stile. The tongue fits into the slot on the rim lock portion which is on the inside. They are suitable for either hand of door. The standard type is 92 mm long, but a 60 mm pattern is available for narrow stiles.

Knobsets

These are comparatively modern locking devices which have integral locks and furniture. The mortised section which houses the lock is tubular in shape, but the lock mechanism is in the cylindrical case between the knobs. They are key operated from one or both sides and suitable for doors of 35 mm up to 47 mm thickness.

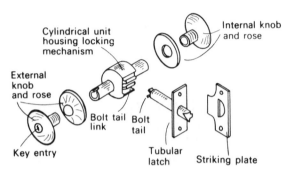

Lockable knobset

Door bolts

Additional security is achieved by fitting bolts to doors. Such bolts are made in a wide variety of sizes and materials to suit every requirement. The three most commonly used are:

1   Straight bolts
2   Necked bolts
3   Flush bolts

Straight bolts

These are entirely face fixed by screws into the rail of the door. The keep into which the shoot passes is screwed to the frame. Available in brass, bronze and aluminium.

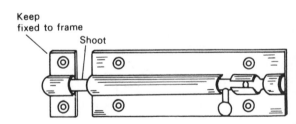

Straight barrel bolt

Necked bolts

This type has a cranked shoot and is designed for the inside face of doors which open outwards.

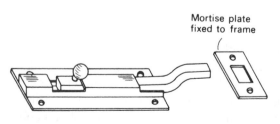

Necked bolt

Flush bolts

As the name suggests, this type of bolt is cut into the stile of the door finishing flush with the face. For this reason they have a neater appearance. Once again a wide range of sizes and finishes are available.

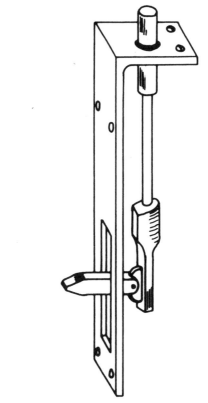

Lever action flush bolt

18 Fixing and fixing devices

Screws and nails, with their respective uses, were described in the previous chapter and it could be argued that much of what will appear in this section could also come under the heading of ironmongery. However, it is felt that special attention should be drawn to these fixing methods so that they will only be used in the situations for which they are intended or designed, thus getting maximum value from them.

Fixing into solid materials

Pallet fixings

This is an old and well tried method of gaining a good fixing. A specially designed *plugging tool* is used to clear the mortar joint between the brick courses in positions required. The pallet is then cut from sawn timber, about 75 × 18 mm, and the end is shaped as shown: an axe is the best tool for this operation. Next, the pallet (which is sometimes called a wood slip) is driven in as tight as possible after which it should be cut off flush with the brick face.

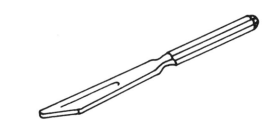

Steel plugging chisel

Hardened fixing pins

These are made from high tensile steel with a bright zinc coating to protect against corrosion.

They are designed for fixing into bricks, concrete and, in some cases, steel sections. Two types of head are available, one is the *butt head* (flat) and the other is a slightly curved head called the *mushroom head*. Butt heads are preferred because they give maximum hold and are also more suitable for cartridge operated tools.

Care should be taken to ensure that the maximum penetration into the actual brick or concrete is 18 mm. Any attempt to drive deeper than this is likely to be counter productive and reduce the holding efficiency. Lengths of pin range from 18 mm up to 100 mm.

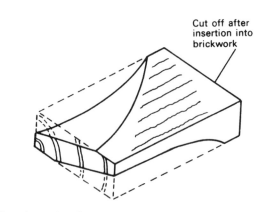

Cut off after insertion into brickwork

The shaped pallet

Mushroom head drive pin

Butt head drive pin

Rawlplugs

Rawlplugs provide an excellent method of screw fixing into brickwork or masonry. They

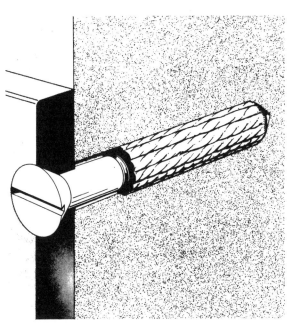

Fibre Rawlplug

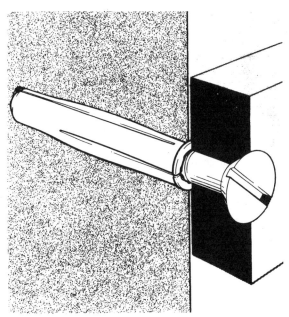

Plastic Rawlplug

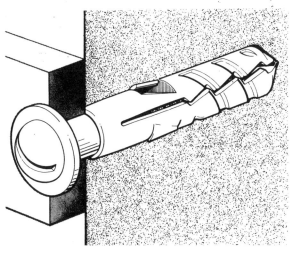

Nylon Rawlplug

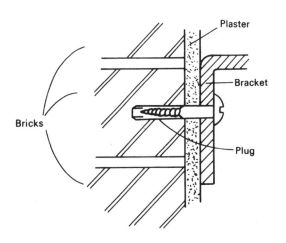

Best position for plug

are made from natural fibre, plastic (polypropylene) or nylon. The method of fixing is based on the principle of expansion of the plug within the hole when the screw is inserted. This means that the plug is *only as strong as the materials surrounding it*, which makes it

dangerous to insert a plug into loose or fracturing material. For this reason, when fixing into brickwork the plug should be situated firmly into the brick itself (not the mortar joint) and as far from the edges of the brick as possible.

As far as the fibre plugs are concerned, plugs of the same gauge number of the screw to be used are made and it is bad practice to use plugs of smaller or larger size than the screw itself. Plastic and nylon plugs are made to suit groups of three gauge sizes of screw, e.g. green (plastic) plugs are suitable for gauge numbers 4, 6 and 8 while the M5 (nylon) type will be suitable for 6, 7 and 8 gauge screws.

Hole forming

The importance of holes being clean cut and of the correct diameter cannot be overstated. Two common methods of forming the holes are as follows:

1 *Rawldrill and toolholder*
When the hole has to be formed by hand methods this is an ideal tool. Most sets have various sizes of drill to suit most purposes and an ejector tool to remove the drill from the holder.

The tool (with correct size drill) is held in position and struck firmly, but not too heavily, with a hammer. The tool should be turned between each blow to clear masonry dust and to prevent jamming.

Goggles should *always* be worn for eye protection when using this tool.

2 *Masonry drills*
These drills can be used in a hand wheel brace, but are more often used with a powered palm grip drill running at its slower speed.

The drill has the appearance of an ordinary drill, but is tipped with *durium* which is a very hard, abrasion resistant carbide. The fluting is designed to allow the masonry dust to clear quickly and avoid clogging. A standard series is made with sizes from 4 mm to 25 mm.

A firm direct pressure behind the drill should be applied when drilling and eye protection should be worn.

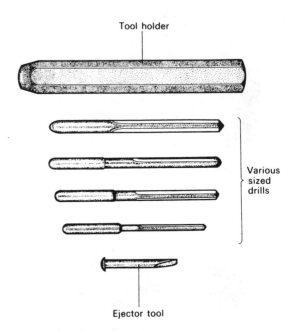

Tool holder

Various sized drills

Ejector tool

Rawldrills with toolholder

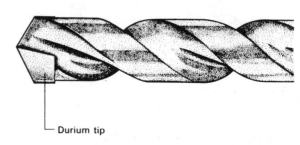

Durium tip

Durium tipped masonry drill

The use of rotary percussion drills with masonry fixing has also been mentioned in Chapter 3.

Rawlbolts

These fixings are specifically designed for heavy duty work. They are made with an iron shell held together by a steel ferrule at one end and a spring wire clip at the other.

When an expander is drawn up the inside of the shell by the tightening effect of the bolt the segments are forced apart to exert a tremendous grip on the surrounding masonry. The bolts are also made with a thread projection where required.

Holes of the correct diameter are formed with a rotary percussion tool and durium tipped drill. Because of the pressure exerted by the rawlbolt, these holes should be kept away from the edge of the masonry by a minimum distance equal to five times the diameter of the bolt being used. It is advisable to follow the manufacturer's instructions as regards size and distances apart of Rawlbolts. These are available in table form.

Rawl ramset

Cartridge operated tools were referred to earlier in connection with hardened nails. The Rawl ramset is an integrated cartridge-powered fixing system.

As with most other power tools they have

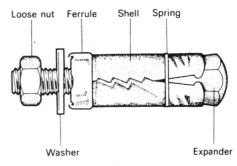

Projecting bolt type

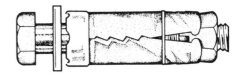

Loose bolt type

Types of Rawlbolt

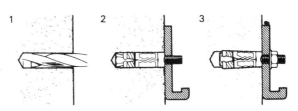

1 Drill a hole of recommended diameter and depth

2 Insert the bolt and position fixture on the projecting bolt end

3 Apply the nut and washer and tighten to expand the shell and secure the fixing

Sequence of operation using Rawlbolts

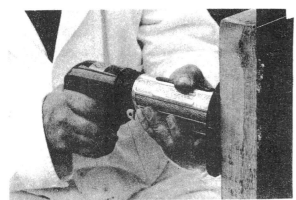

Cartridge tool being used to fix timber to a steel column

design features that comply with BS 4078 which ensures safety in use as well as maximum efficiency. However, it is wise for such tools to be used *only* after a course of instruction for this type of tool has been satisfactorily completed. The tools are capable of securing good fixings into steel, concrete or other masonry. A full range of hardened nails and threaded studs are also available for various uses which includes battening, skirtings and partitioning.

Fixing into hollow materials

All the foregoing methods of fixing are only suitable when fixing into solid materials. Today, however, more and more building materials are being used which are of hollow construction. For these materials methods based on expansions within the hole are not suitable because there is not enough solid support for the fixing. Therefore, special fixings must be used in these situations and the following types have proven themselves in many difficult circumstances and given satisfactory service.

Rawlnuts

These consist of a bonded rubber sleeve having a nut bonded in one end and a moulded external flange at the other. When the metal threaded screw is inserted through the flanged end the nut is tightened, thus causing the sleeve to compress itself into a strong demountable rivet.

The rawlnut is suitable for making fixings into plasterboard, cellular and hollow building blocks etc.

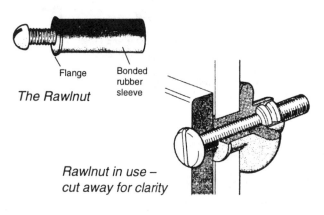

Flange   Bonded rubber sleeve

The Rawlnut

Rawlnut in use – cut away for clarity

Poly-toggle

Also designed for securing fixtures to plasterboard and other sheet material claddings, these are ideal for fixing bathroom or kitchen fittings of medium size. Once installed they will remain in place, enabling redecoration etc. to be carried out. They are suitable for use where the cladding plus fitting thickness does not exceed 38 mm.

Spring toggles

These are ideal for fixing into cavities or with sheet materials. The spring toggle consists of a steel spring-loaded toggle bar which forces the wings to spread when inserted into the cavity. When the screw is tightened the wings are pulled tight against the material, thus causing the load to be spread over a wide area and producing a firm fixing.

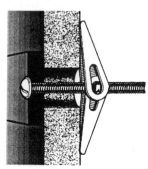

Spring toggle

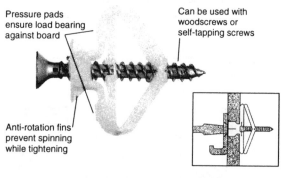

Pressure pads ensure load bearing against board

Can be used with woodscrews or self-tapping screws

Anti-rotation fins prevent spinning while tightening

Screw must be 25mm longer than the sum of the fixture thickness and the plasterboard.

Poly-toggle

Self-drill fixing

This method of fixing to plasterboard is strictly for lightweight fittings. A drill is not needed because the fixing will drill itself into the board.

Available in nylon or metal, they have the advantage of being able to be removed and re-used if required.

Frame fixing

In situations where a window or door frame has to be fixed into an existing opening or where a frame has become loose in its opening this fixing provides a simple and secure method of achieving a satisfactory solution. The 8 mm or 10 mm hole (depending on the size needed) is drilled straight through the frame into the brick, stone or blockwork in as many positions as considered necessary to achieve a dependable fixing.

Interset

These are designed for use in materials where access to the reverse side is not possible. It can be used in similar circumstances to the toggle fixings. When tightened, the screw causes the thin metal 'legs' to fold at pre-determined positions resulting in an efficient fixing with considerable hold.

A 10 mm hole must first be drilled through the material into the cavity. The interset is then inserted so that the teeth on the flange penetrate the material. When the screw is tightened the legs fold back onto the reverse side of the material.

Precision drill tip geometry for faster cutting

Flange prevents accidental push through in plasterboard

Faster Flute Helix means less torque and fewer turns required

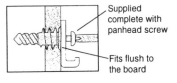

Place fixture in position, insert screw and tighten until secure. Use only with screw provided

Supplied complete with panhead screw

Fits flush to the board

Self-drill fixing

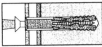

Extended zig-zag design for enhanced expansion

Pozi-head screw provided, suitable for use with screw caps

Durable PA6 grade nylon for extended fixing life. Tough nylon sleeve prevents frame distortion during tightening

Frame fixing

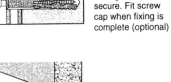

Insert the screw and tighten until secure. Fit screw cap when fixing is complete (optional)

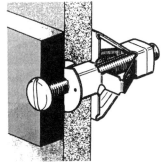

Interset

Questions

1 What type of nail would be used to fix the following:
   (a) rafters and joists,
   (b) skirting to a block wall,
   (c) an architrave around a door frame,
   (d) a small moulding to a panel.

2 List the types of screw which have a decorative use.

3 What are the uses of a domed head screw?

4 What is a pilot hole for?

5 List the advantages gained when using:
   (a) rising butts,
   (b) parliament hinges.

6 Why is it bad practice to insert Rawlplugs into mortar joints?

7 What device would you use to fix a panel to cavity construction in:
   (a) a ceiling,
   (b) a wall.

19 Site setting-out and levelling

Before any work on site can commence the setting-out and levelling must be done. This entails the correct and accurate positioning of the foundation so that the construction of the building can proceed. This part of the work is carried out with levelling instruments of fine accuracy and other equipment which can be prepared or made up by the carpenter as required.

Setting-out equipment

The following pieces of equipment are used to accurately set-out a site.

Pegs

Pegs are usually made from 50 × 50 mm sawn softwood 600 mm long and pointed at one end to make the task of driving them in easier. The upper half should be painted white for easy recognition.

They are driven into the ground in various positions on site to denote levels and important positions at the setting-out stage.

Profiles

These consist of 100 or 150 mm boards, 25 mm thick, nailed horizontally across two stakes driven into the ground. Trench widths and positions of brickwork are marked on the profiles, usually by a small saw kerf (cut), to

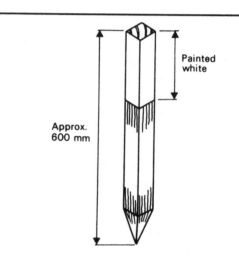

Typical peg made from 50 × 50 mm softwood

Plain profile showing brick and concrete positions

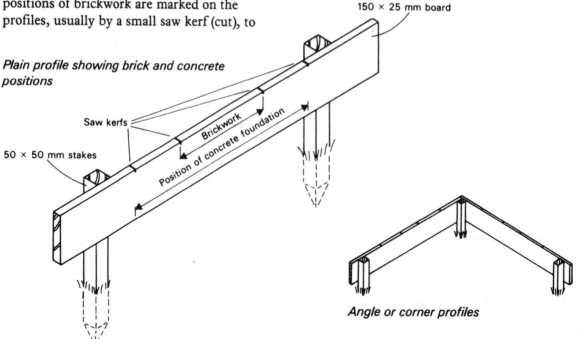

Angle or corner profiles

prevent them being accidently erased. Profiles are used at quoins (corners) and at each end of a trench where foundation concrete is required. Their length varies, but 1.200 m profiles are adequate for most purposes. Plain or angled profiles (for corners) may be used.

Boning rods

It is important that boning rods should be of good quality timber and accurately made. The rod is made from 80 × 18 mm wrot timber, jointed to form a 'T' shape. The cross members are normally about 300 mm long, but these can vary. The stem is between 1 m and 1.5 m long. Boning rods are used in conjunction with levelling instruments or by themselves in determining levels and trench bottoms etc. in the initial stages of work on site.

Levelling board

This must be accurately edge-planed parallel and have at least one true straight edge. This

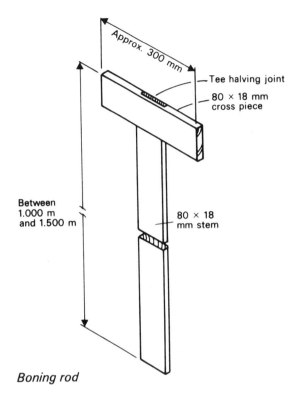

Boning rod

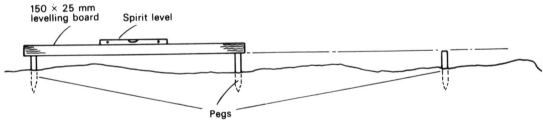

Transferring a level using levelling board and spirit level

edge is used with a spirit level for setting in pegs at required levels. The board is normally about 150 × 25 mm and a popular length is 3 m.

Builder's square

This is a large wooden square with a diagonal brace to maintain its accuracy. The builder's square is an important piece of equipment for obtaining accurate right angles. It is worthwhile spending a little time framing this square together with traditional joints at the angle and

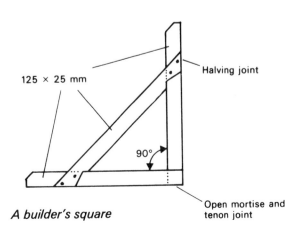

A builder's square

at each end of the brace and gluing it together with a good quality external glue. Made in this way the square will give longer, more accurate service.

Site square

This is an accurate instrument which is primarily for setting-out large buildings. It has a range up to 91.5 m. The head consists of two telescopes which are fixed at right angles to each other. A spirit bubble in the top indicates whether the instrument is set level and ready for use. The instrument can be levelled by adjusting the sliding legs of the tripod. The datum rod must also be accurately positioned before any readings are taken.

Setting-up the site square

1 Set up the tripod, making certain that the bolts are tight.
2 Place tripod with the datum rod immediately over the corner mark.
3 Position the spike on the datum rod and tighten in position.
4 Attach the instrument to the tripod and release the locking screw.
5 Release the tripod leg screws so that the spirit bubble can be adjusted until it is central.
6 Check that tripod leg screws and spike are tightened up and also re-check bubble. Sightings can now be taken.

Cowley level

This is another very easy to use instrument which is accurate to within 5 mm in 30 m. The level is mounted on a tripod and, because of a pivoted mirror arrangement, it will give accurate readings even on uneven ground.

A staff which is marked off in metric units and a target which can be raised or lowered as required are used in conjunction with the level.

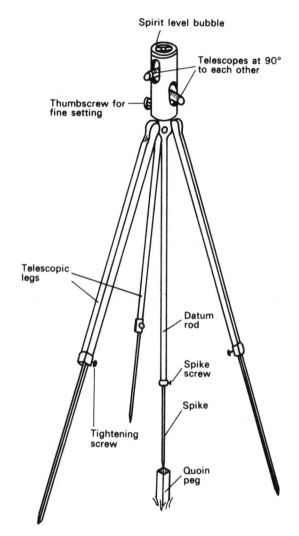

Site square set up on a tripod

To set up the 'Cowley'

1   Spread the tripod legs to give a firm
     standing.
2   Place the instrument on the spike, which
     will release the pivoted mirror, and
     commence sighting.
3   The target is positioned by a second person
     until a level reading is obtained.

Water level

This is a simple yet very useful levelling
instrument based on the principle that water
finds its own level. It consists of two glass or
plastic sight tubes connected by plastic or
rubber tubing up to 30 m in length. The sight
tubes have brass screw caps which are removed
when the level is in use, but tightened down in
position when not in use. The instrument is
particularly useful when level points need to be
transferred round corners, or from inside to the
outside of a building.

To transfer a point the first sight tube is held
close by the given point while the other tube is
held near to the second position. The caps are
then removed and the water surface in the first
tube is held against the given point. The water
surface level in the second tube indicates the
level position.

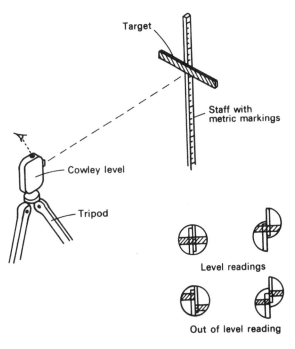

Note: So long as the target meets in the centre,
the split circle can be ignored

The Cowley level

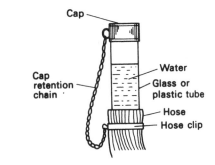

The water level sight tube

Setting-out

In the construction of a normal dwelling the first thing that needs to be determined on site is the *frontage line*. This is the line beyond which the new building must not project. It can be positioned by referring to the layout plan of the building.

Pegs are driven in at positions which are clear of the structure and a string line is then strained between these two pegs (1 and 2), secured to nails in the top of each. Two more pegs (3 and 4) are then driven in on the frontage line in the position of the quoins (corners) of the building. It is from these pegs that a line is strained at 90° to the frontage line. On a single rectangular building of modest size, the builder's square would be adequate to set this right angle. Alternatively the 3–4–5 formula can be applied by means of tape. By measuring 3 m on the frontage line and 4 m on the return with a 5 m hypotenuse a right-angle will have been formed.

By either of these methods the lines are strained from pegs 3 and 4 to two new pegs (5 and 6) which again are kept well clear of the structure. From this point it is possible to position the last two quoin pegs (7 and 8) at distances taken from the drawing between the frontage line and the rear line of the building.

As a matter of habit and in order to check the accuracy of the angles, the squareness of the setting-out should be checked by means of diagonal lines. This is done by measuring diagonally from peg 3 to peg 8 and then from peg 4 to peg 7 — these distances should be exactly the same if the building is rectangular and all angles are right angles.

Having 'pegged-out' the building the profiles are then set up in position. The positions of the trench and brickwork are transferred to the profile boards from the lines which are strained between the pegs when setting-out.

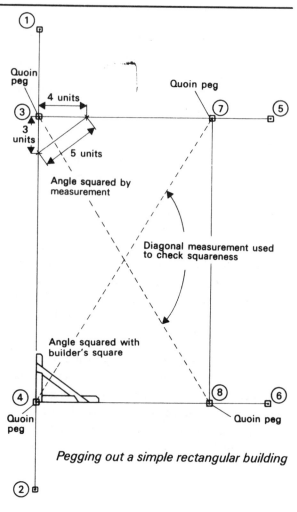

Pegging out a simple rectangular building

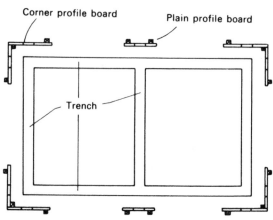

Trenches dug from profile board positions

Levelling

It is when the trenches for foundations are about to be dug that the *site datum* assumes importance. The site datum is a point on the site which governs all other levels of the building and to which all other levels are related. The datum point is fixed at a convenient position and height, normally on some fixture near or on the site such as the top surface of a drainage inspection chamber cover.

If the datum point is off site it should be transferred to a convenient position within the site boundary and marked by a *datum peg*. This peg is accurately positioned and often concreted in to ensure that no movement takes place which may lead to inaccuracy. It is also common-place to construct a small guard around the datum peg to further protect it from damage.

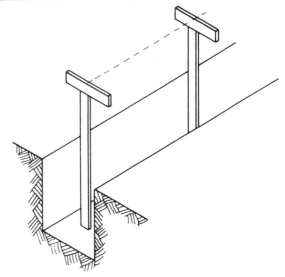

Trench bottom being levelled using boning rods

From this datum point all levels of the building will be set. This includes the drainage runs, which are of paramount importance if they are to link up with the main drainage system accurately and in the position required.

At this stage of the work all measurements and levels should be checked and re-checked. It can be extremely costly if an error is discovered after a great deal of work has been carried out.

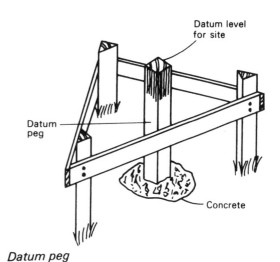

Datum peg

Questions

1 Describe the construction and use of *profiles*.

2 Describe the following instruments and their uses:
   (a) site square,
   (b) Cowley level.

3 Which instrument would you use to transfer a level datum from one room to another? Describe the sequence of operations.

4 Describe two methods of squaring a building when a site square is not available.

20 Access

A range of equipment is available to give access when the work to be carried out is in a difficult position or out of reach from the ground. However, it must be remembered that the risk of accident increases considerably when working from any type of staging, whether it is a simple stepladder or a large scaffold. For this reason the access equipment to be used must be selected on the basis of safety, even if this means taking longer to complete the job. Anyone intending to use any form of access equipment therefore must satisfy themselves according to the following points:

1 Is the equipment appropriate for the work in hand?
2 Is the equipment in good condition and perfectly safe to use?

Work should only proceed when the answer is 'yes' in both cases.

Storage

Access equipment is not usually in continuous use and must be stored between jobs. The importance of correct storage cannot be overstated. Every item must be stored in such a way as to prevent any deterioration, such as distortion, twisting or rusting to any component, which, however minor, will reduce its safety and reliability for future use. An example is shown of a timber ladder correctly stored on brackets, and the distortion that results when it is not.

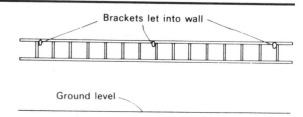

Ladders should be stored off the ground with adequate supports

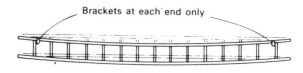

Exaggerated effect of too little support

Maintenance

All access equipment must be maintained to the highest standards of safety. The most effective way to do this is to carry out regular inspections when it is in use *and* while it is in storage. The Regulations state that full scaffolds *must* be inspected at least weekly and, in addition, immediately after a spell of bad weather, e.g. storm, snow. Any item which appears to have developed a fault must be withdrawn at once. Lubrication of all metal parts as well as an occasional coat of exterior quality varnish to wooden members must be an ongoing practice. This will not only maintain safety standards but also contribute to the longer usable life of the equipment.

Ladders

Ladders are used to give access to high work which is relatively light and simple in nature such as painting, minor repairs/replacements. The types most commonly used are made of aluminium alloy or timber (also called Pole Ladders). Aluminium ladders have the advantage of being light as well as strong and, when securely placed, provide a safe means of access. For convenience, extension ladders are used to reach greater heights. When this is so, the user *must* ensure that the recommended overlap of rungs is strictly observed. These are as follows:

1   2-rung overlap for heights up to 4.8 m.
2   3-rung overlap for heights up to 6.0 m.
3   4-rung overlap for any height over 6.0 m.

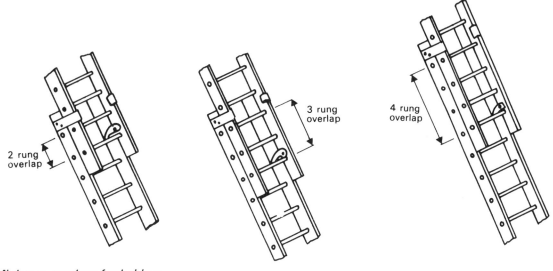

Minimum overlaps for ladders

Pole ladders are mostly used to give access to tubular scaffolds. All ladders should be firmly secured (preferably by rope lashing) at the top and bottom. Where this is not possible a second person should be asked to 'foot' the ladder while it is in use. At no time should more than one person work from the same ladder, and at no time should a ladder be wedged under one side to compensate for uneven ground. The best working angle for a ladder is considered to be 75° which can obtained by setting the foot of the ladder 1 m away from the wall for every 4 m of height, as illustrated.

To comply with British Standards recommendations rung spacings should be 250 mm and be shaped to give maximum foothold to the user. The BS also require that alloy ladders should have a prominent notice warning of the dangers when working near any source of high voltage electrical current such as a railway line or overhead cables.

Step-ladders

These, also, are made from aluminium and timber. They are available in various sizes and are very useful in situations where ladders are too large and for work inside a building. The treads should be at least 90 mm in width with similar spacings as ladders. A frame is hinged to the back which, when fully open, maintains the steps in a rigid and safe working position. Under no circumstances should step-ladders be used unless they are fully open. For timber step-ladders the width of opening is governed by a cord whereas alloy steps are fitted with a locking bar.

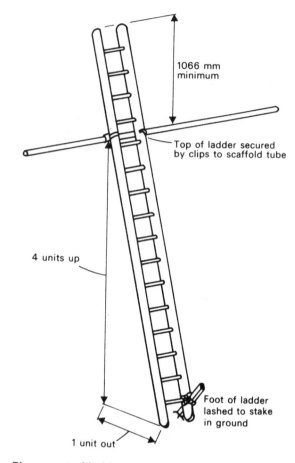

1066 mm minimum

Top of ladder secured by clips to scaffold tube

4 units up

Foot of ladder lashed to stake in ground

1 unit out

Placement of ladder

Trestles

These are used to provide a lightweight staging from which to carry out work at higher levels such as ceilings. They are mostly made from timber and are constructed in such a way as to enable boards or staging to be placed on the cross members (which are spaced 500 mm apart) at the required level. Once again the

trestles must be placed on firm, level ground and must always be used in the fully open position.

Scaffold boards

These are made from softwood which should be straight-grained and free from large knots. They are available in a range of lengths up to 4 m. The most common boards in use are 225 × 38 mm but thinner boards (32 mm) are available for light work and thicker ones (50 mm) for occasions when it is necessary to load the staging with heavier materials such as bricks or blocks. To prevent the ends splitting, the corners of the boards are removed and a galvanised metal strap is secured. 32 mm boards must be supported at 1 m intervals and 38 mm boards every 1.5 m. The end of the board must never overhang its support by more than four times the board's thickness, i.e. 38 mm × 4 = 152 mm (for all practical purposes this figure is rounded to 150 mm).

Scaffold board

Lightweight staging

Sometimes called 'Youngmans' after the company who first designed and marketed them, they provide a very strong and safe platform from which to work if correctly used in conjunction with trestles. Manufactured in timber using tie-rods and reinforcing wires for greater strength, their overall width is 450 mm and a range of lengths up to 7 m are available. Another advantage is that up to three persons can work from them at any time. The rule concerning overhang (thickness × 4) applies as for scaffold boards.

Mobile towers

This type of access is becoming increasingly popular since manufacturers are designing them for greater safety as well as easier assembly. They are aluminium tower systems fitted with hinged working decks for easy access, and lockable castors which combine safety with mobility. Outriggers can be attached to each corner at the base to enable greater working heights to be achieved, but the manufacturer's recommendations **must** be strictly adhered to with regard to these. The safe working load for towers is 275 kg (606 lb) and must not be exceeded. Under no circumstances should steps or any other form of 'hop-up' be used on the working deck in order to gain additional height. If towers are

Components

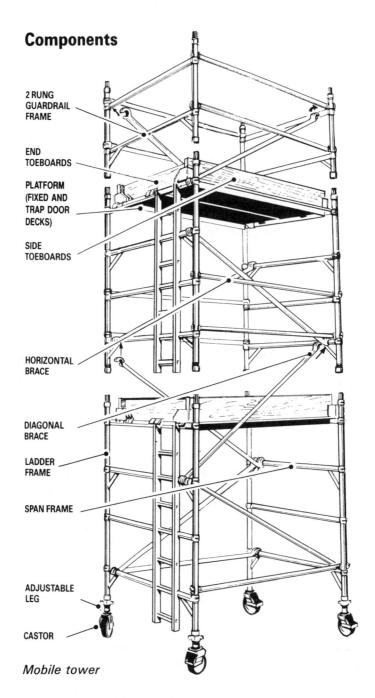

2 RUNG
GUARDRAIL
FRAME

END
TOEBOARDS

PLATFORM
(FIXED AND
TRAP DOOR
DECKS)

SIDE
TOEBOARDS

HORIZONTAL
BRACE

DIAGONAL
BRACE

LADDER
FRAME

SPAN FRAME

ADJUSTABLE
LEG

CASTOR

Mobile tower

being used in the open air, work should be discontinued if the weather is very windy. They must be moved from the base only, using person power. Furthermore, the deck **must** be free from persons and materials during the move. Towers that are more than two-and-a-half times their base measurement in height must *not* be moved in this way. **An initial check must be carried out to ensure that the ground is capable of supporting the tower.**

Tubular scaffolds

These will normally be erected by specialist personnel. The tubes used are of aluminium alloy because of its superior weight/strength ratio and these are linked together using a range of fittings. Only the appropriate fitting should be used in each situation. The range includes couplers, joint-pins, reveal-pins and base-plates. Swivel couplers are non-load-bearing and are used to connect tubes at any angle. *Universal couplers* are load-bearing and

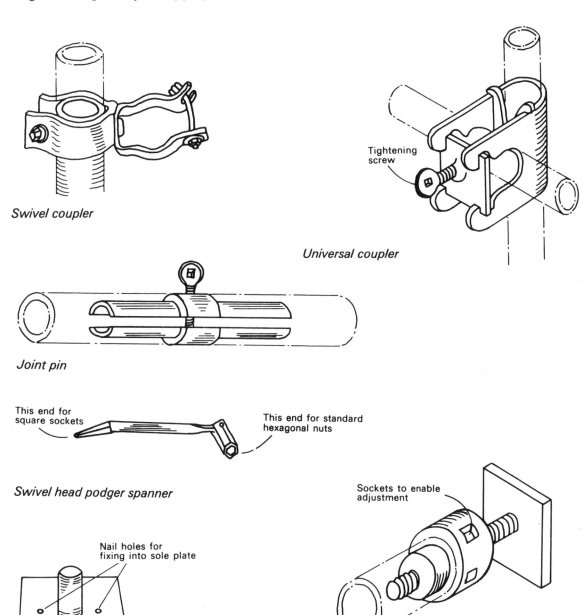

Swivel coupler

Universal coupler

Tightening screw

Joint pin

This end for square sockets

This end for standard hexagonal nuts

Swivel head podger spanner

Nail holes for fixing into sole plate

Sockets to enable adjustment

150 × 150 mm base plate

Reveal pin

are only used to connect tubes at right angles to each other. *Joint-pins* are used for connecting tubes end-to-end for extra length. *Reveal pins* are inserted into the ends of the tubes and tightened to give a rigid horizontal or vertical hold between door or window openings.

Two types of scaffold are used: *Putlog* (for new building) and *Independent* (erected around existing buildings). In the Putlog scaffold the putlog tube has a flattened end to enable it to rest on the new brick courses as they are built. There must be at least 75 mm bearing into the brickwork, as shown, to comply with the safety regulations.

All working platforms must consist of at least four boards and placed at 1.35 m intervals. When the working platform is 2 m or more above the ground a guard rail and toe guard are compulsory. When the working platform is to have bricks or other materials placed upon it the gap between the guard rail and toe guard must be enclosed. This can be adequately achieved using wire mesh or similar material. All scaffolds must be securely tied to the building and be inspected by a competent person at least once a week. In addition, they must be thoroughly inspected following a spell of bad weather, e.g. snow or storm. The safety requirements, as they apply to all scaffolds, are illustrated.

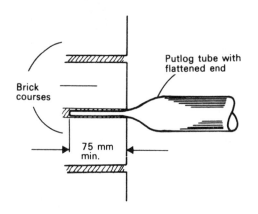

Minimum safe depth of putlog into brick course

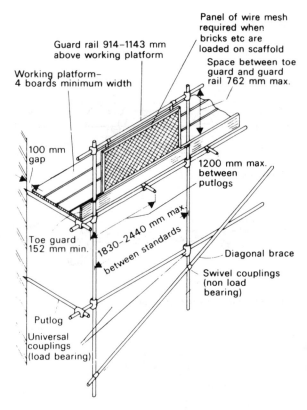

Safety regulations for scaffolds

Questions

1 Why is the storage of equipment so important?

2 What is the safest angle for working on a ladder?

3 State the number of rung overlaps required when using an extension ladder 6 m from the ground.

4 Why should steps and trestles be used only when fully open?

5 What is the maximum overhang permitted on scaffold boards or lightweight staging?

6 What check should be made before using a mobile tower?

7 At what height of scaffold is a toe guard and guard rail compulsory?

8 Name the two types of tubular scaffolds used in construction and describe the difference.

9 What is the minimum number of boards permitted for a working platform?

10 Describe the rules for the inspection of scaffolds.

21 Good working relationships

The development and practice of the skills contained in this book will enable the carpenter and joiner to perform his/her craft to a high level of competence. Maximum efficiency, however, will only be achieved when a conscious and continuous effort is made to contribute to good working relationships. These include:

1 Safety
2 Communication
3 Working with others
4 Problem solving
5 Client relationships

Safety

It is of paramount important that every person engaged in the Construction Industry is 'safety conscious' not only for their own benefit but also for the protection of others. The purpose of the Health and Safety at Work Act (HASAWA) 1974 is to ensure the safety, health and welfare of everyone in their place of work, and it is legally binding on employers and employees alike. The Construction Regulations Handbook is an excellent guide to all the safety and welfare legislation affecting building operations and should be studied by all operatives. It is produced by the Royal Society for the Prevention of Accidents (RoSPA) and it sets out in clear language what is required of everyone. The safety information contained in the various sections of the book have been dealt with at some length and need not be repeated. In addition, however, the Act states that safety notices and posters must be displayed for all to see, particularly in woodworking machine-shops and where hazardous substances are stored. Approved safety signs are now commonplace and the following examples will often be found (their message must be clearly understood and strictly followed):

Category:	Prohibition	Mandatory	Warning	Safe condition
Shape	Circle with crossbar	Solid circle	Triangle	Solid square or oblong
Colour	Red	Blue	Yellow with black border	Green
Meaning	You must *not* do this	You must *do* this	Caution: risk of danger	The safe way: where to go in an emergency

Do not use ladder

Wear eye protection

Risk of fire

Exit — follow arrow

Great emphasis is also placed on the provision and wearing of safety clothing. The following rules should be rigidly observed by all operatives:

1 Hard hats and sturdy boots to be worn at all times.
2 Goggles to be worn when dust, chips or sparks are present.
3 Masks or respirators to be worn where toxic fumes or injurious dust is likely to be breathed in.

Details of all accidents should be recorded in a book kept for the purpose, regardless of how minor they may seem at the time. **NB. If the accident causes the operative to be off work for three days or more it must always be reported.**

No attempt should be made to rectify a potential danger by anyone who is not qualified. Such situations must be reported to the person in charge who will call on the appropriate authority to deal with the hazard. Where the numbers on a site justify it, the operatives are entitled to nominate one of their number to carry out the duties of the Safety Representative whose main task is to advise the management on the implementation of safety policy on-site. Another important function of the Safety Representative is to take all reasonable steps to prevent injury to personnel on-site.

The Health and Safety Executive (HSE) appoint Inspectors who have the authority to enter any site or workplace and check all records, equipment, etc. to ensure that the law is being observed correctly. Inspectors have the power to issue prohibition notices to stop dangerous practices being continued and, ultimately, to prosecute anyone who seriously breaches the law.

In general, all employees have a responsibility to ensure that:

1 They adopt safe working practices at all times for their own and their colleagues' well being.
2 They co-operate with others to enable them to fulfil their health and safety duties.
3 All equipment provided for health and safety purposes, e.g. fire extinguishers, is safeguarded and not interfered with in any way.
4 The fullest use is made of all equipment and facilities intended for health and safety purposes.

For their part, employers have a responsibility to ensure that:

1 A safe working environment is maintained at all times.
2 Training is given in all appropriate cases.
3 The workforce is kept fully informed on all matters of safety.
4 Full co-operation is given to the development and promotion of all health and safety requirements.

Communication

The accuracy with which information is passed from one source (or one person) to another plays a major part in the successful completion of any contract, however large or small. This means that it is absolutely essential to ensure that all information/instruction is as concise and clear as possible to avoid any misunderstandings. The three main methods of communication used in our everyday working lives are:

1 The spoken word
2 The written word
3 Visual aids, e.g. drawings, graphics, photographs, etc.

These can be used singly or in any combination that best suits the specific requirements, bearing in mind the need to avoid complication. The carpenter and joiner will frequently use the written word in making up a cutting list to enable the wood machinist to prepare the materials for a job whilst, at the same time, using a combination of drawing and written words to prepare a rod from which to set out and construct the same job.

In the construction industry communication usually begins with the client giving instructions to the architect who, in turn, produces a complete set of drawings and a specification from which the work will be carried out. Where the work is of a high cost, all this will be done using the standard forms produced by the Royal Institute of British Architects (RIBA) and is, therefore, unlikely to lead to error.

Much greater care needs to be exercised on smaller contracts where there is more likelihood of misunderstanding. The self-employed carpenter and joiner, for instance, cannot afford errors; his work will be keenly priced leaving no room for costly mistakes. In these circumstances there must be a clear understanding before work starts. The client must be left in no doubt about exactly what is being undertaken and the costs involved. This is best done using headed notepaper containing a full, written description and any visual aids that will assist in conveying the full picture.

Accurate communication is also important in the placing of orders as deliveries of wrong goods cause delay and inconvenience. Descriptions of goods required must be made clear to the supplier, using all references in the supplier's catalogue by which the items can be identified. Orders placed by telephone should always be followed by written confirmation. Goods which are delivered to site in an unsatisfactory condition or in the wrong quantity should not be accepted. The reason must be clearly stated on the delivery note and the consignment returned.

Trade journals and other literature are the most common forms of communication used to keep everybody up-to-date with new materials, working practices and trends that are considered to be beneficial to good working relationships.

Working with others

With the growth in the number of self-employed persons and sub-contractors now engaged in the construction industry, there is an ever-growing need for complete co-operation and co-ordination if the highest standards are to be achieved. The various crafts and specialities must not be seen as separate entities; they must be fully integrated. On large contracts there is often someone whose sole responsibility is to oversee the integration of all the different trades. However, where this is not the case, there is a requirement on all workers to have an understanding of allied trades in order that they can carry out their particular work in a way that will assist the others. As far as the carpenter is concerned, this might involve forming access traps in floors and fitting removable panels in other forms of casing to enable electricians and plumbers to gain access to cables, pipework etc.

There will also be many occasions when the

carpenter will need to work with others engaged in the same craft. If the spirit of teamwork is entered into whole-heartedly it will almost certainly result in the work being completed more quickly, more enjoyably, and to a better standard. To make certain that everyone is fully aware of what is required in order to achieve maximum benefits from this policy it is wise, where possible, to arrange co-ordination meetings. These will provide an opportunity for all concerned to agree a mutual approach to the work in hand. The

ability and willingness to work in harmony with others are features that must be developed alongside technical skills if carpenters and joiners wish to make the most of the working opportunities that come their way. Those who are able to offer these skills in addition to a high degree of craft competence will be more eagerly sought after than those who are not. Employers and clients alike will be more willing to engage those who are known to be reliable in this way as a further guarantee of quality assurance.

Problem solving

It is doubtful that there is ever a working day when the carpenter and joiner is not involved in overcoming some sort of problem, however minor. There is not the space available here to analyse *all* the difficulties that are likely to be encountered, but it may be possible to formulate a general approach to what is an important subject. We shall look at the mainly technical problems which are likely to confront the carpenter and joiner, leaving aside the logistical problems which are linked with planning.

The accuracy with which a problem can be diagnosed will depend very largely upon the degree of knowledge possessed by the carpenter and joiner over a wide area, often beyond the confines of craft knowledge. It could be that a tool which is required to perform a certain operation is not functioning correctly. The majority of joiners, for instance, have experienced using a rebate plane which is forming a ragged rebate. If the joiner knows that, to cut clean rebates, the cutting iron must project from the side of the plane by at least 0.5 mm the problem will be instantly recognised and corrected.

Following diagnosis the next step must be to discover the cause and remove it. It is essential that this is done *before* any reinstatement is carried out in order to avoid damage to the new work. A common example of this is a timber ground floor which is under attack by dry rot. Diagnosis will be aided by the musty smell and cuboidal cracking of painted skirting associated with it. Closer investigation reveals blocked air bricks and a damaged DPC as the obvious cause. After ripping out all affected timber and burning it, it is essential to clear all air bricks in order to create proper air circulation, and to repair or renew the DPC *before* installing new timbers.

Experience will enable the carpenter and joiner to solve problems quickly, not least because the vast majority of them occur frequently. However, where this is not the case, a more painstaking process of eliminating the likely causes will have to be followed. No problem should, for expediency, be left unsolved because this could lead to further, perhaps more serious, trouble in the future. If it means that the problem needs to be referred to more specialist personnel, do not hesitate to do so, particularly if there is an inherent danger.

Client relationships

This aspect of the carpenter and joiner's career can often mean the difference between complete success or total failure. From the client's point of view, in addition to being a very competent and skilled operative, the qualified employee must also be seen as a person of honesty and integrity. These attributes create a basis of trust which will lead to all parties being confident that the best value for money has been achieved. This is the case whether the operative is dealing with the client directly or as an employee of a company. If, from the start, the client is treated with courtesy (not patronisingly), a far greater understanding of the requirements of the job will be reached, and at the same time any problems that are envisaged can be discussed. Misunderstanding and disagreement at the conclusion of the work can be avoided in this way thus paving the way for an amicable financial settlement.

Another possible source of irritation to the client is when a particular material or item is not immediately available. Once again the explanation should be made tactfully giving, where possible, helpful suggestions as to alternatives, or procedures to minimise the effect.

There will be occasions when the client will seek advice before deciding to go ahead with a given project. If the carpenter is employed by a contractor, it would be correct procedure to refer the matter, and, even if advice is given, it *must* be on the basis that the final decision rests with the contractor. For the self-employed carpenter, any advice given must be carefully considered, bearing in mind the possibility that other trades may be involved, and this will have implications for the final cost. Here, again, there is the need for complete honesty, and if any doubt exists about the advice being offered it should be made clear at the outset.

Tact and diplomacy should always be observed in any dealings with the client, particularly if the work is being carried out on the client's own premises. Building work is notorious for the dust and disruption it causes. A little extra effort made by the carpenter to minimise this, and simple actions such as making certain that any property which could suffer damage is adequately covered and protected, will go a long way to ensuring that confidence is maintained, leading to a satisfactory conclusion.

Attempts by the carpenter and joiner to avoid any situation that may cause friction will, without doubt, lead to a much more harmonious relationship with the client.

Questions

1 What is the purpose of the Health and Safety at Work Act?

2 What is contained in the Construction Regulations Handbook?

3 Describe the shape and meaning of four safety signs.

4 What is the main task of the Safety Representative?

5 Describe two responsibilities of each of the following under the Health and Safety at Work Act:
(a) Employee
(b) Employer

6 How are operatives kept up to date with new practices/materials etc. in the industry?

7 Describe the advantages to be gained in successfully working with others.

8 List the sequence of actions to be followed in solving a problem.

9 Why should courtesy be maintained at all times in client relationships?

Index